Arthur Lismer, 1885–1969
Launching the Seaplane
Print, 53.4 cm x 71.0 cm
Canadian War Museum, 8379

CANADA'S AIR FORCES
1914–1999

**BRERETON GREENHOUS
HUGH A. HALLIDAY**

CANADA'S AIR FORCES
1914–1999

ART GLOBAL

Canadian Cataloguing in Publication Data

Greenhous, Brereton, 1929–
Canada's air forces, 1914–1999
ISBN 2-920718-72-X

1. Canada. Royal Canadian Air Force — History. 2. Canada. Canadian Armed Forces — History. 3. Airplanes — Canada — History. 4. Canada. Royal Canadian Air Force — Pictorial works. 5. Canada. Canadian Armed Forces — Pictorial works. 6. Airplanes — Canada — Pictorial works. I. Halliday, Hugh A., 1940– . II. Title.

UG635.C2G73 1999   358.4'00971   C98-941553-8

Publisher: Ara Kermoyan

Project coordinators: Stephen J. Harris and D.S.C. Mackay

Editor: Jane Broderick

Jacket illustration: Detail inspired from a painting by Robert W. Bradford, 1923–
   Armstrong Whitworth Siskin III A, 1977
   Acrylic on canvas
   45.5 cm x 56.0 cm
   National Aviation Museum

Published by Editions Art Global and the Department of National Defence
in co-operation with the Department of Public Works and Government Services Canada.
All rights reserved. No part of this publication may be reproduced, transmitted, in any form or by any means, electronic, mechanical, photocopying, recording, or otherwise, or stored in a retrieval system, without the prior written permission of the Minister of Public Works and Government Services, Ottawa, Ontario K1A 0S5 Canada.

Art Global
384 Laurier Avenue West
Montreal, Quebec H2V 2K7 Canada

ISBN 2-920718-72-X

© Her Majesty the Queen in Right of Canada, 1999
Catalogue number: D2-118/1998E

Printed and bound in Canada

Cet ouvrage a été publié simultanément en français sous le titre de :
L'Aviation militaire canadienne 1914-1999
ISBN 2-920718-71-1

Canada

Art Global acknowledges the financial support of the Government of Canada, through the Book Publishing Industry Development Program, for its publishing activities.

TABLE OF CONTENTS

ACKNOWLEDGEMENTS
9

PREFACE
11

CHAPTER I
13
THE FIRST THREE CANADIAN AIR FORCES

CHAPTER II
25
'BUSH PILOTS IN UNIFORM'

CHAPTER III
43
TRAINING FOR WAR

CHAPTER IV
57
APPRENTICE WARRIORS, 1939–1942

CHAPTER V
87
JOURNEYMEN CAMPAIGNERS, 1942–1943

CHAPTER VI
103
MASTERS OF THE AIR, 1943–1945

CHAPTER VII
121
FROM WAR TO PEACE TO COLD WAR

CHAPTER VIII
139
THE NEW AIR FORCES

HUGH A. HALLIDAY was born in Manitoba in 1940. He holds a Bachelor of Arts degree from the University of Manitoba and a Master of Arts degree from Carleton University in Ottawa. An historian with the RCAF and Canadian Armed Forces (1961–68) and a teacher at Niagara College of Applied Arts and Technology (1968–74), he subsequently joined the staff of the Canadian War Museum and held several appointments, including Curator of War Art, Curator of Photographs and 20th Century Historian, before retiring in 1995. He is the author of several books on Canadian military aviation, including *The Tumbling Sky* (1978) and *Typhoon and Tempest: The Canadian Story* (1992), and contributes regularly to *Canadian Military History* and the *Journal of the Canadian Aviation Historical Society*.

BRERETON GREENHOUS was born and raised in England and came to Canada in 1958. He earned a Bachelor of Arts degree in history from Carleton University in Ottawa and a Master's degree from Queen's University in Kingston. After a brief and unfulfilling spell of university teaching, he joined the Department of National Defence's Directorate of History (now the Directorate of History and Heritage) in 1971 and happily worked there until his retirement in 1996. He is the author or co-author of a dozen books on military history, including the third volume in the Official History of the RCAF, *The Crucible of War, 1939–1945* (1994), and two earlier books in this commemorative series, *The Battle of Vimy Ridge* and *Dieppe, Dieppe,* both published in 1992.

ACKNOWLEDGEMENTS

The Minister of National Defence, the Deputy Minister of National Defence and the Chief of the Defence Staff approved the publication of this book in the summer of 1998 and have provided unstinting support to see it through to completion.

Historians are indebted to their predecessors; writers owe much to their peers. The late Wing Commander Fred Hitchins and the late Flight Lieutenant Fred Hatch strove to preserve and promote air force history in its leanest years; their contributions are found throughout the text. Similarly, Tim Dubé and Peter Robertson of the National Archives of Canada gave full meaning to the term 'public servant,' while Larry Milberry of Canav Books also generously assisted the authors. Bob Bradford was particularly kind in giving us permission to publish details inspired from his painting of the Siskin aerobatic team as well as the complete work in the special colour section (pages 72 and 73).

The publication of this history would not have been possible without the support of the Chief of the Air Staff, Lieutenant General D. N. Kinsman, or that of Dr. G. J. A. S. Bernier and the staff of the Directorate of History and Heritage, National Defence Headquarters.

Wherever practicable, the words of actual participants in the events described, whether in memoirs or interviews, have been used; these are distinguished by italic type.

PREFACE

This book is an important contribution to the marking of the seventy-fifth anniversary, on 1 April 1999, of the Royal Canadian Air Force. In this diamond jubilee year, the men and women of Canada's present-day air force will recognize, celebrate and pay tribute to all their predecessors, who, from the very beginning of Canadian military aviation in 1914, served Canada in peace and war with selflessness, great courage and unflinching dedication.

Beginning with the First World War and continuing through the establishment of the Royal Canadian Air Force as a separate military service in 1924, the inter-war years, the cataclysm of the Second World War, the Korean War, the decades of the Cold War, and the Gulf War, up to the present day, men and women of many nationalities have served in Canada's air force and the military air services that preceded it. In a similar vein, many Canadians have served, principally in wartime, in the military air services of our allies, primarily the United Kingdom and the United States; their contributions, unhappily, cannot be recorded here to the extent that they merit.

No matter where and with whom, these men and women have all served voluntarily, and in wartime they have fought valiantly in the causes of freedom, human dignity and international peace and stability. Tens of thousands have lost their lives in such service and countless others have suffered grievous injuries. Canada should take great pride in her sons and daughters who have fallen and in all those who have served. This book is a tribute to them.

This volume is also intended to recognize and focus attention on those who serve in Canada's air force today. Whether at home or abroad, in the service of the North Atlantic Treaty Organization, the North American Aerospace Defence Command or the United Nations Organization, or in responding to natural disasters, search and rescue situations or other circumstances of human suffering, the men and women of Canada's air force will continue to meet the high standards of excellence and professionalism that have been the hallmarks of Canadian airmen from the outset. They will also be ready to fight, if called upon to do so, in defence of Canada and the cause of freedom, just as those who served before them did.

This book is an eminently readable and thoroughly enjoyable history of Canada's air force — a rich and colourful history that will appeal to all those who are in any way moved by or interested in military aviation. I am pleased and proud to take this opportunity, on behalf of the men and women of Canada's air force, to thank and commend all those who have contributed to the success of this book and to wish its readers many hours of enjoyment.

Per Ardua Ad Astra

D. N. Kinsman
Lieutenant-General
Commander of Air Command
and Chief of the Air Staff

Chapter I

THE FIRST THREE CANADIAN AIR FORCES

The first Canadian to fly in a powered, heavier-than-air machine, on 12 March 1908 at Hammondsport, New York, was Frederick Walker ('Casey') Baldwin, a twenty-six-year-old Torontonian who was one of the original members of Alexander Graham Bell's Aerial Experiment Association. Two months later the AEA moved to Bell's summer home, at Baddeck, Nova Scotia, where local guinea-pig John McCurdy made the first Canadian flight, aboard the *Silver Dart,* on 23 February 1909.*

Even before McCurdy's pioneering flight, two visionary staff officers at Militia Headquarters in Ottawa, Major G.S. Maunsell and Colonel R.W. Rutherford, were showing an interest in military aviation. As a result of their lobbying, in May 1909 the Militia Council invited Baldwin and McCurdy (who together, upon the demise of the AEA, had just formed their own business, the Canadian Aerodrome Company) to demonstrate their product at the militia's summer camp site at Petawawa, Ontario. Arriving at Petawawa with their two 'aerodromes' — aircraft were called aerodromes in 1909 — Baldwin asked for a shed in which to house them, and Militia Headquarters approved the expenditure of five dollars for laths and tarpaper to build it. 'This lavish investment represented Canada's first official expenditure in connection with military flying.'[1]

The subsequent trials were impressive — any flight was impressive in those halcyon days — but not particularly successful. The machines were as fragile as butterflies, with open frames — they could hardly be called fuselages — of wooden spars and fabric-covered wings and tailplanes braced with piano wire. Flying them was an extraordinarily difficult exercise. 'In practically no other acquired accomplishment has man to keep so many groups of antagonistic muscles in a state of static wakefulness, or to perform such variety of constant co-ordinated

---

* Four years younger than Baldwin, McCurdy lived to see the dawn of the space age. He died in 1961, just two months after the Soviet Union's Yuri Gagarin became the first man to fly in space.

leg and arm movements,' wrote an early specialist in aviation medicine. 'The successful flier must be one who has power to co-ordinate his limb muscles with a beautiful degree of refinement....'[2]

The *Silver Dart*, with McCurdy at the controls, made three successful 'hops' of a kilometre or so before crash-landing with irreparable damage. 'If she can hop like that, she can fly,' concluded the army's perceptive quartermaster-general, Brigadier-General Donald MacDonald. Their other machine, the *Baddeck I*, was wrecked on its second flight, when McCurdy stalled while thirty feet in the air. Again the pilot was not hurt, but the trials were over. In Ottawa, the deputy minister of militia, Colonel Eugène Fiset, explained to the *Evening Citizen*:

> You cannot expect a young country like Canada to strike out and adopt a military aeroplane policy....
>
> Who knows what these aeroplanes can do?... Can they lift a great weight? What protection would the canvas planes offer? I think they must find something of a more stable nature than canvas to cover the great wings with. We must wait a great many years yet and experiment much more before the true use of these machines can be demonstrated.

* * *

The first Canadian to fly in uniform — he wore the khaki serge of the Canadian Expeditionary Force — was twenty-one-year-old Lieutenant William Sharpe, of Prescott, Ontario, who had earned his pilot's licence from the American Aero Club in January 1914 and was scratching a living from exhibition flights in the Chicago area when the First World War began in early August. During the second week in September, Sharpe was merely one among 30,000 volunteers for the CEF gathered at Valcartier, Quebec, where his ability to 'co-ordinate his limb muscles with a beautiful degree of refinement' must have seemed of limited value. However, at some point his path crossed that of Edward Janney, a confidence trickster of sterling calibre from Galt, Ontario, who had just wheedled out of the minister of militia an appointment to command the Canadian Aviation Corps.

The CAC was Janney's fabrication, sponsored by the erratic and easily misled Sam Hughes, who had displayed, until he met Janney, a firm aversion to any idea of Canadian military aviation. There is no evidence that Janney ever held a pilot's licence (there was then no legal requirement for one in Canada) but he did have a marvellously sweet tongue — sweet enough to talk Hughes into creating the CAC with a stroke of his pen, commissioning him on the spot, and authorizing the purchase of 'one biplane, with necessary accessories, entailing an expenditure of not more than $5,000.00' — a thousand-fold increase on the earlier disbursement made at Petawawa.

Having enrolled Sharpe in his new command, Captain Janney hurried south, to Marblehead, Massachusetts, where he managed to buy a well-used Burgess-Dunne aeroplane from the company that built them. It was a swept-wing, tailless machine — modern in concept, if not in design and construction — with rudders at the wingtips, two seats set in tandem on the lower wing, and a 35-hp engine driving a pusher propeller. Moved by rail to Lake Champlain and fitted with floats, it took off for Valcartier on 21 September 1914 with a company pilot, Clifford Webster, at the controls. The flight was not an easy one, according to Webster.

> *Started from Isle La Motte at 7:14 AM on Monday and climbed steadily to about 1,500 feet at the time we crossed the border. I don't think anyone shot at us. From there to Sorel our altitude varied from 1,000 to 2,000 feet and the wind, which had been following at first, swung around to our port beam and so proved detrimental. Janney drove about a fourth of the time. At the end of an hour and fifty-five minutes we reached Sorel and I decided to land as the gas was getting low.*
>
> *We managed to get ashore alright and get some gasoline and oil ordered. We started out again at a quarter of twelve and made good time to Three Rivers, where I expected to land, but Janney signalled me to continue. About ten minutes later I heard a knock in the motor, so came down and ran ashore at Champlain (in*

flight 45 minutes). As the motor would still turn over fairly freely, Janney decided (against my advice) to load it up with oil and chance it for Quebec. I managed to get off and get started down river, but the motor was turning over so slowly that I had to keep damn busy to keep up. I did not get above 15 feet at any time and was often only a few inches clear.

After 18 minutes of this inferno the motor gave up the ghost and I landed with regret but not without relief also. After we had drifted for some twenty minutes, a motor boat saw our distress signal and towed us in here [to Deschaillons]³.

The Burgess people promptly dispatched a mechanic by train from Marblehead, and he managed to repair the decrepit engine in a week. The two fliers reached Quebec City on 29 September, barely in time for the machine to be pulled apart and loaded aboard one of the transports carrying the First Contingent to England, wings and engine boxed in the hold and the fuselage on deck.

While Janney was away, Sharpe had recruited a mechanic from the ranks of the CEF's 16th Battalion. Harry Farr, a pre-war British immigrant to Victoria, British Columbia, was working in San Francisco as chief motor mechanic of the Christofferson Aircraft Company when the war began, but had left his job and returned to Victoria to enlist in the Gordon Highlanders.⁴ With the acquisition of Farr, smartly promoted to the rank of staff sergeant, the inchoate CAC reached its maximum strength of three.

The First Contingent of the CEF, including the CAC and the Burgess-Dunne, arrived on England's Salisbury Plain during October, but parts of the aircraft had been damaged during the voyage; it was still in pieces, and may not even have been all in one place. For the moment, however, Janney had more important things to do than arrange for repairs and re-construction. He was busy devising an initial establishment for the CAC, consisting of a flight of four machines (with another four as spares) and a staff of one commander, three pilots, three observers, seven sergeants and thirty-two mechanics, together with five horses and ten trucks.

The start-up cost for the first year, he estimated, would be $116,679.25 — the price of a Canadian air force was growing by leaps and bounds. Janney's calculations inspired the Contingent's British commander, Lieutenant-General Sir Edwin Alderson, to cable Ottawa:

> Two individuals...have accompanied Canadian headquarters from Canada claiming to be aviators authorized by minister but not mentioned in militia orders.... Please cable instructions as to their status, pay, and whether large expense necessary to organize an efficient unit of one flight is authorized.

Hughes, always unwilling to admit his mistakes, responded that the two officers (everyone was ignoring Farr, who presumably was still carried on the pay lists of his old battalion) were to be paid on the same basis as equivalent ranks in Britain's Royal Flying Corps. However, he added, it was not intended to organize a flight; one machine was enough.

In mid-November Janney had drawn an advance from the paymaster for travelling expenses and set off to tour Britain, visiting Royal Flying Corps stations allegedly in search of a suitable site for the CAC's training. When several weeks had elapsed without any further word from him, Janney was marked down as absent without leave. Finally, late in December, when Militia Headquarters in Ottawa learned (from whom, one wonders?) that he was planning to return to Canada, 'intending without authority to give exhibition flights to raise funds to equip complete flight unit for Contingent.' Hughes lost his patience — which was never very great — and decided to sever Janney's connection with the militia. He was struck off CEF strength on 23 January 1915 and thereupon faded from the record until September 1918, when he would (as we shall see) briefly re-appear in a naval uniform.

Sharpe, meanwhile, had begun further training with No 3 (Reserve) Squadron of the RFC, at Shoreham in Sussex. He was killed on 4 February, when he crashed in the machine he had taken up on his first solo flight overseas. That was the end of the Canadian Aviation Corps. 'The body of Canada's first military aviator and first air casualty was brought back to Canada for interment at Prescott...' Staff Sergeant Farr was discharged from the CEF in

May, 'in consequence of Flying Corps being disbanded,' and thereafter joined the Royal Naval Air Service, ending the war with a Distinguished Service Cross and a Distinguished Flying Cross. As for the Burgess-Dunne, it never flew again. Sometime in 1916 its metal remnants — the engine, some hinges and a few other fittings, for the wood and canvas had rotted away — were 'piled in a railroad truck with a lot of other scrap iron and sent to Salisbury to be sold.' *Sic transit gloria mundi.*

* * *

The government showed a distinct reluctance to create a successor to the CAC. There were a number of Canadians who either already knew how to fly and sought commissions or were keen to learn the former and earn the latter — but there was a definite lack of enthusiasm in political and military circles. Prime minister Sir Robert Borden, who apparently was never told anything about the CAC, had remarked as early as October 1914 that his government did not think it desirable to organize an air service during the war; Hughes lost any interest he may have had in aviation upon the demise of the CAC; and the British officers serving as Chief of the General Staff and Director of the Naval Service (who both preferred to see Canadians join the British flying services) were equally negative.

In March 1916, John McCurdy, by now an executive of the American-based Curtiss Aeroplane Company and managing its Toronto flying school, proposed that the government might establish an aviation branch for flying training which could 'easily be later moulded into a permanent branch of the service.' His suggestion was cautiously rejected. Loring Christie, legal adviser to the Department of External Affairs, was asked to formulate a prime ministerial response and did so in classic bureaucratic style.

> ...the conclusion of everybody [concerned] seems to be that the needs of the war are not such as to demand the immediate organization of a distinct Canadian flying service, and that it is therefore better to wait until it has a fair chance of being established on a sound basis when a trained personnel will be available and the conclusions drawn from the experience of the war will have been more carefully considered and formulated.

Against this unsympathetic background, Canadians joined the British flying services, at first in dozens, then in hundreds and eventually by the thousand. Wherever the air war was fought — over England or Germany, the English Channel, the North Sea, the Aegean or the Dardanelles; in Italy, Macedonia, East Africa, Palestine or Mesopotamia; and most of all on the Western Front in France and Flanders — there they were, in ever-growing numbers.

Initially, the RNAS and the RFC took on men who had already learned to fly at Canadian or American schools, while other men paid their own way to England and applied to join there. If they were accepted, the British taught them to fly, as was the case with soldiers already serving with the CEF who were permitted to transfer to Imperial service.* Among the early recruits were four men who would eventually become Chiefs of the Air Staff, two — George Croil and Lloyd Breadner — during the Second World War, and the other two — Robert Leckie and Wilfred Curtis — in the immediate post-war era. Moreover, nearly all those who reached the rank of air commodore or above during the Second World War — the exceptions being, without exception, technical specialists — were recruited during the First World War. It was these men who would give the Royal Canadian Air Force its ethos and identity, and *that* makes their experiences well worth summarizing.

* * *

The war gave an enormous boost to aircraft development. In 1914 the finest military machines could reach speeds of around 100 kph and heights of 3,000 metres or so, the pilot

---

* In 1916 the RFC established a dozen schools of its own in Canada and began training Canadian pilots *in situ;* and once the Americans had entered the war, when deteriorating weather made flying training in Canada impossible, RFC schools moved, lock, stock and barrel, to Texas for the winter.

August 1909: the 'Baddeck' biplane at military trials, Petawawa. The demonstrations were not impressive.
(National Archives of Canada, PA 119432)

Canada's first military aircraft — the Burgess-Dunne, 1914.
(Canadian Forces photograph, RE 17706)

The Canadian Aviation Corps's Burgess-Dunne aircraft coming aboard ship in Quebec, October 1914. (CFP, RE 17705)

making careful turns and shallow dives and sometimes carrying an observer or his weight in bombs, but certainly not both. Four years later, fighters like the Fokker D VII or Sopwith Snipe could travel twice as fast and fly twice as high.* They could be spun and rolled and looped without a wing breaking loose. Unwieldy bombers such as the Handley-Page V 1500 and Staaken R VI could carry a crew of six or more *and* a bombload of 1,500 kgs to targets 800 kilometres away, at a cruising speed of 160 kph.

Tactics and weaponry developed in parallel. At first airmen flew as individuals, the boldest among them using rifles, shotguns or handguns to attack their enemies when the opportunity arose. In April 1915 Second Lieutenant Malcolm Bell-Irving, from Vancouver, was patrolling over the Ypres salient in a Martinsyde Scout. In one of the earliest combat reports ever filed, his commanding officer recorded that:

At about 4.30 pm two machines were observed.... Lt. Bell-Irving attacked these with an automatic revolver, and when all his ammunition was expended did his best to turn them by nose-diving.... Lt. Bell-Irving reports that had he had a machine-gun he ought to have accounted for one...but adds that at 5,500 ft [1,651 metres] *the Martinsyde without machine-gun has to be flown at 65 mph* [102.5 kph] *in order to fly level, and with a machine-gun his speed would be reduced below this. Consequently, unless the hostile faster machine failed to observe his approach (as in fact he did for a considerable time) a machine-gun on a Martinsyde would be of no great assistance.*

For three years the tactical/technological pendulum swung to and fro, giving first one side, then the other, a material advantage. Both sides developed interrupter gears which permitted a machine-gun to fire through the airscrew disc without chopping off the propeller, enabling pilots to aim simply by pointing their aircraft at the target — with the appropriate 'lead' for deflection, if required — and thus encouraging the development of single-seater fighters.

---

* Without oxygen for the pilot or, in the British case, a parachute. Parachutes had been devised and tested, but, it was said, issuing them would make airmen too willing to abandon their machines.

During the spring of 1916 the fearsome 'Fokker scourge' starred the Fokker E-series (the E for *Eindekker,* or monoplane) and the formation tactics introduced by Oswald Boelcke. The RFC retaliated with the Nieuport 17 and Sopwith Pup, and adopted formations of its own.

Before 'bloody April' of 1917, the Germans regained the edge with their Fokker Dr.I (*Dreidekker,* or triplane) and the streamlined Albatros D-series biplanes, both types mounting twin machine-guns; and they exercised the tactical privilege of choosing to either fight or flee. The British responded with the SE 5 and Sopwith Camel but were tactically constrained by their commander's demanding philosophy of *always* engaging in offensive action, usually well over the enemy's lines, a tactic that cost him (and even more his airmen) dearly.

In the end, however, it became a question of whose formations were the largest, whose aircraft were the fastest and most manoeuvrable, whose machine-guns were the most plentiful. In late 1917 Captain G.C.O. Usborne of Arnprior, Ontario, who had been with the RFC since 1915 and was then a flight commander on the Western Front, bluntly explained that *'it isn't a case of pluck and clever flying, it's simply a case of a man who can climb the fastest and dive the fastest.'*

A related insight comes from the experience of a neophyte observer in March 1918, as he helped stem the last great German advance of the war.

The air was so packed with aeroplanes that sardines in comparison seemed to be lolling in luxury. The cloud ceiling was low, about 2,000 feet [600 metres], *and in that narrow space hundreds of machines swooped and zoomed, spitting fire at each other and at the troops below.*

Nickle, my Toronto pilot, dived on German troops marching along a road, machine-gunning them furiously through the airscrew, and as he turned to regain height I continued with my gun.... Aeroplanes flashed by on all sides, friend and foe almost impossible to distinguish.

We dropped our bombs on a German [artillery] battery....

One hundred and seventeen individual combats were recorded on that day by the RFC, with thirty-seven of them being identified as 'decisive' victories.

Strategic bombing had been practised, sporadically and ineffectively, from September 1914 when RNAS aircraft described as 'old veteran servants of the Crown' flew from Antwerp in a vain endeavour to thwart zeppelin attacks on England by bombing the airship sheds at Düsseldorf. Only one of four reached the target, its six 8-kg bombs failing to explode because they were dropped from too low a height. In 1917 the Independent (Bombing) Force was set up to attack German factories and generating stations in the Ruhr from French bases; but when the war ended civilian morale was becoming the target of choice and plans were afoot for Handley-Page V 1500s to raid Berlin from England.

Other specialities arose, from artillery co-operation to strategic reconnaissance to ground support, each of them with their own particular problems, machines and equipment, and each becoming more sophisticated as the war progressed.

* * *

Sam Hughes had been driven from office in November 1916. Five months later Sir Robert Borden, thoroughly frustrated by his recent visit to England, where he had encountered a British reluctance to recognize the importance of Canada's contribution to the war, cabled his high commissioner in London, Sir George Perley, expressing a new-found view:

> Since returning to Canada representations have been placed before me which indicate that Canadians in Flying Service are not receiving reasonably fair play or adequate recognition. There seems to have been a disposition from the first to assign them to subordinate positions and to sink their identity. They were forbidden to wear any distinguishing badge to indicate that they were Canadians. They have been discriminated against in promotion.... The question of establishing a Canadian Flying Corps demands immediate...consideration.... I am inclined to believe that the time for organizing an independent Canadian Air Service has come and that we must ask the Imperial authorities to release all Canadians now in the British Flying Service.

Shifting *all* the Canadians in British service into a national formation would have had drastic implications. When Borden cabled, in May 1917, more than a quarter of the airmen on the Western Front were Canadians, and in another year that proportion would have risen to a third. At the very least, the British would have been left with too many chiefs and not enough Indians!

Nor were all Borden's criticisms entirely justified. In London, Sir David Henderson, the commander of the RFC, dipped and dodged, demonstrating statistically that the absence of Canadians in the higher ranks was logical and their lack of recognition comprehensible, if not reasonable. Borden's 'inclination' would soon weaken. However, a seed had been sown and his change of front, although it would prove to be uncertain when political push came to shove, established a climate in which the efforts of others would finally bear fruit.

The proximate cause of the first fruiting, however — creation of a Royal Canadian Naval Air Service — was the imminent spread of German U-boat activity into the western Atlantic. In the fall of 1917 a new class of U-boat easily capable of transatlantic voyages began to enter service, with the *Kriegsmarine*, and the British Admiralty warned Ottawa that 'an attack by one of the new enemy submarine cruisers might be expected in Canadian waters any time after March [1918].'[5]

U-boats were still too few in number and their endurance too limited to pose a major threat in the open ocean, where merchantmen were difficult to locate without the assistance of electronic intelligence — a benefit that would only accrue in a later war. But in the approaches to ports and the constricted waters of estuaries, bays and straits, where shipping inevitably clustered, submarine warfare could be immensely profitable. In such circumstances aircraft had already proved their value (in conjunction with the convoy system) in forcing submarines to remain submerged, where their speed was much lower and their endurance even more limited. If U-boats were to reach across the Atlantic, to lie off Halifax, Sydney or Saint John, or to intrude into the Cabot Strait and Gulf of St. Lawrence, an air arm would be a key part of essential defensive measures.

The Americans (who had entered the war in April 1917) were equally threatened, but their navy already had an air arm and bases where most needed. However, in April 1918 a conference in Washington concluded that additional stations should be established at Dartmouth and Sydney as soon as possible and that subsequently outposts should be added on Sable Island, at Canso and in the Magdalen Islands.

Setting up a naval air service would take time as well as money, and until their neighbours could spread their own wings the United States agreed to supply all the necessary flying equipment and personnel while Canada recruited and trained its own aircrew and mechanics. The Americans began to arrive in early August and four Curtiss flying-boats were quickly assembled at the order of Lieutenant Richard E. Byrd, Officer-in-Charge, US Naval Air Force in Canada.* The machines made their first flight over Halifax on 25 August, but no one had bothered to inform the authorities at the Citadel.

Considerable excitement has been reported to me arising out of the unexpected appearance of the air service machines yesterday. No information has reached us regarding the addition of this service to the garrison. This I would be glad to get as the fortress is equipped with anti-aircraft defences.

More flying-boats arrived, and were supplemented with a flight of seaplanes. Escorts were provided for convoys leaving and nearing port but they never did locate a U-boat, although the three that sailed into North American waters managed to sink 110,000 tons of shipping between Cape Hatteras and Newfoundland during the last two months of the war.

Meanwhile, the Royal Canadian Naval Air Service was created by order-in-council on 5 September 1918, with an establishment for ninety commissioned aircrew and up to one thousand ratings — one recruit being Sub-Lieutenant E.L. Janney! But, said PC 2154, 'the proposed organization shall be regarded as temporary for the purpose of meeting the needs of the war'; and seventy-two days later, on 13 November 1918, just two days after the war ended, the British officer in charge was notified that the RCNAS had 'ceased to exist.'

Closing down the RCNAS meant the end of all their hopes for those who had enlisted in the new arm; their services were no longer required. Janney, for one, sank without a trace, perhaps to practise — one likes to think — bigger and better scams on less wary prey in more salubrious climes. However, three weeks later the government had apparently begun to doubt the wisdom of its decision and the minister of the Naval Service, C.C. Ballantyne, explained to his deputy minister, G.J. Desbarats, that 'the RCNAS is not abolished, and the action that is now being taken is only until such time as the Government decides on the details and policy of a permanent Air Service.' Perhaps with that in mind, Ottawa happily accepted the donation of twelve flying-boats, twenty-six spare engines and four kite balloons by the Americans when they left.

Shortly after the RCNAS had been formed, another Canadian air arm had been established in England, and much of the credit for this second service must go to Sir Richard Turner, the nationally minded businessman-turned-general who had been 'kicked upstairs' to command Canadian forces in Britain after his failure at St. Eloi, in April 1916. Turner, intrigued by the idea of a Canadian air force overseas, had calculated in June 1917 that there were at least 1,200 Canadians then serving in the RFC and the RNAS. As time would tell, those figures were a gross under-estimate. However, in July he passed the list to Sir George Perley, the Cabinet minister stationed in London to take charge of the Overseas Ministry that Borden had established in the wake of Sam Hughes. Turner's covering letter was strongly phrased for official correspondence:

> I consider that as Canada is supplying such a proportion of the personnel, that we should proceed with the organization of a Canadian Flying Corps. This would enable the Canadians to take their rightful place in the Imperial

---
* Byrd would later win fame as a polar aviator and explorer.

William Frederick Nelson Sharpe (Prescott, Ontario) of the Canadian Aviation Corps. Sharpe was killed in a training accident.
(CFP, PL 39933)

Proposed cap badge for the Royal Canadian Naval Air Service, 1918, incorporating aerial, naval and Canadian iconography.
(CFP, CN 3932)

Sopwith Dolphins, Canadian Air Force, England, early 1919.
(NAC, PA 6024)

Forces, to receive the full credit for the work being done by them, and to provide an organization of experienced personnel to carry on the flying service after the war.

I would propose the Canadian squadrons be organized as rapidly as conditions would allow with the ultimate object of having, if possible, a Canadian Brigade in the Field, together with the necessary reserve formations in this country and in Canada.

In September Turner was at it again:

There continues to be a large and insistent demand for Canadian personnel, and we are sending a large number of our best young men to fill these demands....

I feel very strongly that we should at once proceed with the organization of a Canadian Flying Corps.... I feel that it is humiliating that a nation taking such a share in the war as is Canada, should not have an organization in this arm of the Service when she is supplying such an important part of the manpower in the Imperial Services.

But there was a practical problem which had not occurred to Turner, guaranteed to frighten off any Canadian politician. It was pointed out by Wing Commander Redford H. Mulock, the highest-ranking Canadian in the RNAS:

The efficiency...in the Field of any Flying Organization depends almost entirely upon the machines with which it is equipped.... The supply of machines is absolutely controlled by the [British] Air Board, and any Canadian organization would be absolutely subject to it in regard to the supply required. The Canadian Authorities, however, would be held responsible to the people at home for any misadventure which might occur to the personnel on account of inferior equipment, and yet...it would be a matter of great difficulty to ensure that such Canadian Flying Units as might be organized and placed on active service would receive their just proportion of the most up-to-date and efficient equipment which is placed in service.

Mulock's memorandum discouraged, in turn, overseas minister Perley, minister of militia Sir Edward Kemp and Prime Minister Borden. Turner could only respond with the most careful analysis yet of Canadians in the RAF, on 1 April 1918 — the day it was formed by amalgamating the RFC and RNAS. According to his new figures, which were probably reasonably correct,* until that time 13,345 Canadians had served in the British flying services and 10,990 remained on strength.

Turner buttressed his case for a Canadian air force with supporting letters from a number of prominent Canadians serving overseas, including Lieutenant-General Sir Arthur Currie, commander of the Canadian Corps in France, Brigadier-General A.C. Critchley, the highest-ranking Canadian in the RAF, and Major W.A. Bishop, then the only Canadian airman to have been awarded a VC and generally recognized as the premier RAF fighter pilot. Canadian newspapers, too, had taken up the case for a national air force with enthusiasm. 'We have now more men in the air service overseas than we had all told in the South African War,' noted the Toronto *Star* on 25 May 1918, 'yet they do not form a Canadian contingent.'

The British, understandably, desired to delay the inevitable. 'Concrete action on these lines,' Sir William Weir, the secretary of state for air, wrote to Kemp on 4 June 1918, '...should be avoided for the present, and I would suggest to you that the principle should be to work out a scheme which could be brought into force, say, next winter.'

Neither Perley nor Turner was willing to wait, although Borden might well have been. Despite the foot-dragging of the RAF representatives, who succeeded in severely limiting the number of Canadian squadrons to be formed, the next day a meeting between them and Brigadier-General H.F. McDonald, Turner's senior staff officer, and Thomas Gibson, the Overseas Ministry's assistant deputy minister, resolved

---

* There was no legal definition of Canadian and no category of Canadian citizenship at that time, and exact figures were therefore impossible to calculate. Those born in Canada obviously qualified, but how long did an immigrant have to live in the country before he could be called a Canadian? We do not know Turner's criteria.

that 'the formation of two Canadian Air Squadrons shall be proceeded with forthwith.' It was not the eight-squadron wing that Turner had proposed but it was better than nothing, while the British could easily spare the thirty or forty aircrew involved without crippling the RAF.

The British were to be responsible for supplying aircraft and equipment and for providing training facilities for replacements while the squadrons were in Britain or a theatre of war; the Canadians would enlist the men and pay them. On 19 September 1918 an order-in-council issuing from Ottawa confirmed this arrangement, but only for 'the purposes of the present war.' That qualifier made it clear that no decision had yet been made in Ottawa on the question of a permanent Canadian air force.

No 1 (Fighter) Squadron, equipped with Sopwith Dolphins, was placed under the command of Major A.E. McKeever, who had distinguished himself with the RFC as a Bristol Fighter pilot on the Western Front. No 2 (Day Bomber) Squadron, equipped with the DeHavilland 9a, was given to Major W.B. Lawson, an ex-RNAS bomber pilot who had most recently been flying Handley-Page 0/400s from French bases with the Independent [Bombing] Force. Neither the Dolphin nor the DH 9 was well thought of by those most qualified to know, thus elevating to fact the possibility that Redford Mulock had foreseen. McKeever would have preferred Sopwith Snipes; Lawson, Bristol Fighters.

The selection of flying personnel was not completed until 18 November, when the war had been over for a week. Initially it had been intended to crew both squadrons with a mixture of decorated veterans and pilots fresh out of flying school, but in the event nearly all of those chosen had outstanding records — MCs, DSCs and DFCs were two a penny and DSOs were not uncommon. By any measure these would be élite squadrons. However:

> For battle-hardened pilots the transition to peacetime conditions, and to the very difficult challenges of organization and administration, was difficult enough. The task of formulating a policy that would persuade the government to maintain the CAF was one for which most of them were not equipped.

There were indeed considerable administrative challenges. Two hundred and thirty-seven other ranks had volunteered for ground-crew training from CEF depots in England* and been sent to RAF schools to be taught their new jobs. All too often, and with some justice, they resented RAF standards of discipline and messing, resulting in several minor mutinies. 'The senior N[on] C[ommissioned] O[fficer]s, who held their positions because of technical competence, had no experience in handling men, nor had many of the flying officers.' Matters were eventually straightened out, however, by returning the worst offenders to their former units and posting in a CEF sergeant-major to establish Canadian-style discipline.

In order to improve the overall administration, Colonel C.C. St. P. de Dombasle was brought in to command the CAF (his title was soon changed to Director of the Air Service — a projection of what was to come!), the two squadrons were designated a wing, and another well-decorated veteran, Lieutenant-Colonel Robert Leckie, was appointed wing commander.

* * *

The war was over, but there was no word on the fate of the CAF. 'Is it the intention to use these two Squadrons as the nucleus of the Canadian Royal Air Force?' asked Major McKeever in December 1918. In March, a joint RAF-CAF conference at the Air Ministry considered the matter and drafted a paper reflecting the views of airmen who were already firm advocates of strategic bombing. They argued that 'a complex new art of aerial strategy, as different in its application from either Naval or Military, as these two are now different from each other,' needed to be developed.

It was probably a mistake to put the case for a permanent air force in such terms. As the one dissenter — a British officer — wondered,

---

* The RAF eventually discovered 234 Canadian mechanics already in their ranks and willing to transfer, but by then the CAF had begun training these men from the CEF.

what was the relevance to Canada of arguments based upon the strategic implications of airpower? Canada was 'hardly open to aerial attack except from the U.S. and I understand her policy does not contemplate armed resistance to that country.' On 30 May, with that policy paper in front of them, the Cabinet in Ottawa decided that Canada did not need, and could not afford, a military air force. Sir Edward Kemp, still the minister of militia, thereupon told the prime minister, 'After we have demobilized...it will be apparent to members of Council that a great mistake has been made.' He then wrote to Gibson, expressing the hope that 'a less elaborate organization and one which would readily adapt itself to peace conditions in Canada may be worked out.' And that, in fact, was the route that Canadian airpower would take in the coming years.

There was to be a parting gift, some recognition by the British of the part that Canadians had played in the world's first air war. Canada already had the twelve flying-boats donated by the Americans; now the British added, from their war surplus stores in the hope that Canada would still maintain an air force, two more flying-boats, sixty-two Avro 504K trainers, twenty-two DH 4 and DH 9 bombers, twelve SE 5s and two Bristol Fighters, and a solitary Sopwith Snipe, together with a quantity of tools, engines and other equipment. Canada also inherited the RFC training stations in Ontario, lock, stock and barrel. If the government should choose to establish another air force in the not-too-distant future, the requisite materials would be at hand.

Chapter II

'BUSH PILOTS IN UNIFORM'

The end of the First World War put hundreds of surplus aircraft on the market and let loose enough airmen anxious to continue flying, for either pleasure or business, to make some sort of overarching government control essential. At a minimum, aircraft and pilots needed to be registered and licensed, airfields and meteorological services established, and air traffic regulated.

Ottawa's interim solution was to prescribe an Air Board to do these things, although the Cabinet dogsbody — minister without portfolio A.K. Maclean — assigned to introduce the enabling bill into Parliament, made it quite clear that:

> Speaking generally, the Government has no settled policy in respect to aeronautics.... It is open to the Government to embark upon various forms of aerial activities, such as the conveyance of the mails, the survey of forests, fire protection, patrols by the Mounted Police, and so forth, but at the present time the Government has reached no definite policy in respect of any of these matters.[1]

As well as dealing with the essential problems of civil aviation, the Air Board looked at the need for a military air force. In February 1920 its members observed that 'if Canada is again obliged to engage in war, it will be necessary for her to rely upon an air force as well as upon land and sea forces; and the longer the period intervening before the commencement of such a war, the greater will probably be the importance of the air force by comparison with these other forces.' They then concluded:

> ...that war strength in the air must ultimately depend upon civil or commercial air strength; that most of the members of a war air force must normally pursue peaceful occupations (preferably but not necessarily, in connection with air navigation), that war formations should exist only upon paper and not in the form of embodied units, and that war

training should be periodic, intensive and widespread....

In peace even the instructional and administrative personnel at the training stations should consequently, with the fewest exceptions, be civilians temporarily assuming military duty. It is obvious that this would result in a lower peace efficiency than if a more numerous permanent professional military personnel were relied upon, but peace efficiency is not the primary consideration. A war organization so constructed as to be comparatively inefficient in peace but reasonably efficient in war is very greatly to be preferred to a war organization which shows a high degree of efficiency in peace but breaks down when it is called upon for war service.[2]

The last clause of the first paragraph quoted above was as absurd as the last sentence of the second paragraph was astute. Indeed, one must wonder at the wisdom of three Cabinet ministers (including the minister of militia and defence), two deputy ministers and the judge-advocate-general of the militia who could collectively believe that a competent air force might exist without some form of 'embodied unit,' however small, to set standards by example.

Nevertheless, on 18 February 1920 the government approved the formation of a non-permanent, largely part-time, Canadian Air Force as one of several aspects of the Air Board. Before the end of June, a headquarters staff of six had been appointed, under the command of Air Commodore A.K. Tylee, who had most recently been inspector of training for RAF Canada. The CAF was assigned a budget of $800,000 for the first year, and stations with 'caretaker' permanent staffs were to be established at Camp Borden, Ottawa, Dartmouth, Vancouver, Winnipeg and High River, Alberta. A rather stylish dark-blue uniform was devised, with buttons and badges of silver, and authority granted to recruit 1,340 officers and 3,905 airmen. By the end of 1921, 1,281 officers had been enlisted, but only 1,350 other ranks. A very small proportion of these officers and men, some fliers, some groundcrew and some cooks and bottle-washers, were on indefinite full-time duty; but most would be called to the colours only for one month of training every second year.

The CAF's first assignment, in line with the Air Board's overall policy to further civil operations, was an experiment in fire patrols and aerial survey of the areas north and west of Lake St. John, Quebec, carried out at the request — and expense — of the provincial government. Two HS-2L flying-boats, part of the American post-war gift, were flown in from Halifax to Roberval in July 1920, and 'flying continued uninterruptedly until the end of October when winter put an end to the work for the season.' The province was pleased with the result and requested further work for 1921; it was prepared to increase expenditures 'considerably.'

Another innovation, in September, saw an HS-2L (flown by Captain C.M. McEwen, who will re-appear later in this story) sent to Haileybury, on Lake Timiskaming, to spend eighteen flying hours surveying the effects of the spruce budworm on northern Ontario forests. Out west, a fourth HS-2L carried a Dominion entomologist to the Northwest Territories/Alberta border so that he could inspect, and photograph from the air, swamps where malarial mosquitoes were believed to be breeding. At the end of the year a nod was given to military activities when the same machine was flown to Vancouver Island 'for coastal defence reconnaissance.'

Flying training — really refresher training, since all of the CAF's pilots were veterans was concentrated at Camp Borden, north of Toronto. The machines used were Avro 504Ks from the British post-war gift. With costs amounting to about six dollars an hour (what a shocking price!), only eighty-six officers were called out that year, and they averaged 11.23 hours flying time each. Mostly, it was a matter of individual practice, with some aerobatics and formation flying. However, at Petawawa two DH4 hours were devoted to artillery co-operation — 'spotting' for militia gunners carrying out their annual practice — and an aerial survey of the camp. In western Canada, a DH4 from High River made an artillery co-operation flight of about four hours' duration over Sarcee Camp.

Service training in the use of flying-boats was gained 'by a special detachment of the CAF

Bristol F.2b, Avro 504 and DH9 at Rockcliffe, 1920: a cross section of the aircraft received from Britain as part of the Imperial gift. (NAC, HC 25)

Major R.A. Logan, Captain A.L. Cuffe, Major J.A. Glen, Major L.S. Breadner, Major A.E Godfrey. The distinctive Canadian Air Force uniform is apparent. (CFP, RE 14542)

Swinging a propeller on an Avro 504K for the instruction of trainees, Camp Borden, 1927. (CFP, RE 15833)

mobilized for a week-long exercise with the RCN at Halifax.' The enterprise concluded with one machine flying eighty kilometres off-shore to rendezvous with the fleet and escort it back to Halifax, 'all the while maintaining wireless telephonic communication with the base, to which messages were also transported by carrier pigeons released at sea.'

That was to be the pattern for the next three years, until the RCAF was created in 1924; and for the eleven years after that, until 1936, about three quarters of all flying time was spent on civil operations and one quarter on flying training, with no more than a wink towards military matters. Those who flew in the CAF and RCAF between the wars were simply *'bush pilots in uniform,'* according to one of the originals, Air Vice-Marshal T.A. Lawrence.[3]

* * *

Between 1921 and 1924, however, the CAF evolved from a branch of the nominally civilian Air Board to a quasi-independent air force, determinedly military in organization and style even if essentially civilian in function. Its metamorphosis into the RCAF was the result of second thoughts about the need for at least one 'embodied unit' — a mutation probably made easier by the fact that in 1922 the Air Board became part of a new Department of National Defence, together with the Militia and the RCN.

Thus the CAF became a subsidiary of the Canadian forces, and in February 1923 King George V approved of the prefix 'Royal' being added to the organization's title. That same year saw the introduction of *ab initio* pilot training — it was unreasonable to assume that the war veterans would be able to fly forever. Nine cadets were inducted; four graduated, but two of them would be killed in flying accidents within two weeks of each other in the summer of 1927 (flying was still a hazardous business, even without the dangers of war) and one resigned his commission later that year. The remaining graduate, Pilot Officer C.R. Slemon, would become chief of the air staff in 1953 and deputy commander of NORAD (North American Air Defence Command) in 1957. His recollections of the RCAF's early years simply confirm Lawrence's aphorism about bush pilots in uniform. *'I never thought of a weapon; I never saw a weapon or fired a machine-gun, or whatever. We were just as busy as could be, doing purely civil government flying. ...all along there were military elements, but they were tiny in comparison to the civil government air operations.'*[4]

It took a year for Ottawa to complete drafting new regulations and orders for the RCAF, replacing those devised in 1920, but by 1 April 1924 everything was ready, and on that day — the sixth anniversary of the creation of Britain's Royal Air Force — the Royal Canadian Air Force came into existence. The RCAF was a virtual copy of the RAF in style, structure and equipment, even down to the pattern and changed colour of the uniform, which became a much lighter grey-blue. Unlike the RAF, however, the RCAF was not an entirely independent service although its subservience was more apparent than real. It was led and administered by a director — the first one was Wing Commander W.G. Barker, VC, DSO and Bar, MC and two Bars, etc., etc., then (and now) Canada's most decorated veteran — who was responsible to the militia's chief of staff; and not until 1938 would it be granted full equality, with its own chief of staff.

The RCAF, moreover, was still not much concerned with training for war. Forestry patrols, with particular emphasis on fire prevention, continued to be the largest item in their work load. In 1924 a project to photograph all of Canada from the air began, and that usually took second place in RCAF priorities, with fisheries protection an easy third. At a time when alcohol purchase and consumption without a doctor's prescription was still illegal in Ontario, New Brunswick, Nova Scotia and Prince Edward Island,* one intriguing by-product of the photo-survey was a frantic shifting of illicit stills as 'moonshiners' feared that their operations were being exposed from the air.

---

* Prohibition had been introduced nationwide in March 1918 as a wartime measure. Quebec rejected it in 1919 and British Columbia and the Yukon in 1920, followed by Manitoba in 1923 and Alberta and Saskatchewan in 1924. It lasted until 1927 in Ontario and New Brunswick, until 1930 in Nova Scotia and until 1948 in PEI.

Major R.A. Logan displays the CAF ensign at Craig Harbour, Ellesmere Island, August 1922, while investigating possible use of aircraft in the Arctic. (CFP, RE 13996)

Captain Clifford M. McEwen, former fighter pilot and future Air Vice-Marshal, in 1922 during forestry patrol work, northern Ontario. (CFP, RE 16625)

During the RCAF's first year there were seven exercises involving the militia or navy, amounting to 'about 50 hours of flying' out of a total of 3,941 hours. In 1925, seventy-three hours of military training were recorded out of a total 5,111 flown. Early that year, Flight Lieutenant J.L.M. White, DFC and Bar, who had had a short but brilliant career as a fighter pilot on the Western Front in 1918, and Flying Officer R. Cross, the first RCAF airman to be commissioned from the ranks, became the new service's first flying casualties, killed in a mid-air collision. As a result, parachutes were issued and training in their use became obligatory — although it was not until 1929 that the first emergency use of one saved an airman's life, since nearly all flying accidents occurred during take-offs and landings when the height necessary to make a parachute descent practical was lacking.

A notable milestone came in 1927–28, when the government required accurate information about ice and weather conditions in the Hudson Strait if Churchill, Manitoba, was to fulfil its potential as a port for moving prairie grain to Europe.* A detachment under the command of Squadron Leader T.A. Lawrence — his was an acting rank, granted for the duration of the task — was assigned to the job and six single-engined Fokker high-wing monoplanes equipped with floats and skis were purchased especially for the mission. (These machines were never taken on RCAF strength, however, and were given civil registrations.) The pilot sat in an open cockpit, with his crew and/or passengers in an enclosed cabin behind him.

The Fokkers, together with six pilots, twelve fitters and riggers (fitters were aero-engine mechanics; riggers were responsible for the airframes) and eighteen months' supplies, were moved north by ship and deposited on the shores of the Strait in July 1927. Along almost a thousand kilometres of shoreline three bases were established: at Port Burwell, on Nottingham Island and at Wakeham Bay. Two each of aircraft, pilots and groundcrew were stationed at each base, together with a medical officer, radio engineer, two other ranks from the Royal Canadian Corps of Signals, an RCMP constable, a storekeeper and a cook.

Aircrews were made up of a pilot, flight engineer — usually a fitter, as engine problems were the likeliest source of trouble — and *'an Eskimo as a handyman and guide in the event of a forced landing. The Eskimos proved very adaptable crewmen on all occasions: worth their weight in gold.'* Piled in the tiny cabin, together with the engineer and the Inuit helper, were a primus stove and fuel, enough food for at least ten days, a tool kit and insulated engine cover, an inflatable raft, distress signals, a rifle and ammunition, sleeping bags, ropes and axes; with the addition of snowshoes, blow torches, ice axes and snow knife for the winter months.

During the expedition's ten months of arctic flying there were three occasions involving forced landings. The first, caused by heavy fog, was uneventful and meant no more than an overnight camp on the sea ice. The second, brought about by *'converging snow storms,'* kept the expedition leader, Lawrence, and his crew ice-bound for ten days, but they came to no harm.

The storm broke as we got the aircraft tied down and the oil drained. We were close to land where the Eskimo built an igloo which became our home for the next eight days, during which time we seldom saw the aircraft. Under very cold but clear and calm weather, we spent one full day digging out and cleaning snow out of the cockpit and from the engine. The snow had penetrated every crack and chink. We got the engine started with only a few hours to spare on the 10th day, took off roughly but without incident, and arrived at Wakeham as darkness set in.[5]

The third occasion threatened a more serious fate when a blizzard caused Flying Officer A.A. Lewis, flying out of Port Burwell:

...to let down within a few feet of the ice pack where accurate navigation became well-nigh impossible, what with the local magnetic disturbances, the oscillations of the compass needle, the extreme turbulence of the air and almost zero visibility....

Suddenly I saw, immediately below, what appeared to be a stretch of clear, greenish ice,

---

* The railway to Churchill was completed in 1929 and substantial quantities of grain were subsequently exported through the port until the 1960s, but the route never did realize its full potential.

Vickers Vicking G-CYET about to embark on an aerial survey in northern Manitoba, July 1924. Note the equipment carried, including a paddle. (NAC, FC 193)

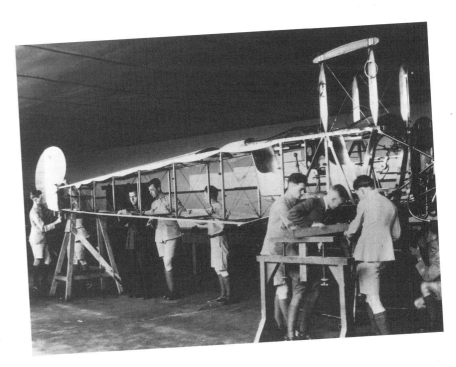

Technical training at Camp Borden using an Avro 504 airframe. (CFP, RE 16264)

Fokker Universal G-CAHH during conversion from skies to floats, Hudson Strait Expedition, May 1928. (NAC, PA 055481)

and fervently praying that it extended ahead at least a short distance I cut the engine, whereupon the aircraft dropped like a stone almost vertically.... When we hit the ice, so strong was the wind that the aircraft stopped almost immediately but the [ice-]pinnacles were so numerous that we could not avoid hitting one head-on, and the aircraft finished up with the tail up in the air and the nose and skis buried in a deep snowdrift against an ice-pinnacle.

No one was hurt, but were they in Ungava Bay or somewhere on the North Atlantic Ocean? Lewis had been allowing a twenty-degree correction for drift, which should have placed them in the Bay. *'I couldn't reconcile in my mind the fact that any gale would require a greater correction than that.'*

Their Inuit helper, Anaktok, built an igloo for the night and the next day they started walking east, into the storm, in poor visibility, at a temperature in the vicinity of forty degrees below zero! However, by nightfall Lewis was beginning to wonder whether they were heading in the right direction. The second morning dawned bright and clear.

We were completely surrounded by tightly packed ice-pinnacles and could not immediately sight anything on the horizon, so climbing to the top of one of the pinnacles I was almost afraid to look towards the west, for if I saw nothing it would, without doubt, dash all our hopes for our survival. However, when I at last glanced to the west, there, clearly etched in the sky, were what appeared to me to be mountain peaks white with snow. I could scarcely contain my joy.... A rough calculation convinced me that we must be about 50 miles [80 kilometres] out in the Atlantic, for taking into consideration the height of the mountains...on a clear day, one has a range of vision of approximately 50 miles. [6]

They reversed course and, after seven long days, walked off the ice on to the Labrador coast about a hundred kilometres south of Port Burwell. There they were found by a passing Inuit hunter and taken to Burwell, Lewis and his crewman suffering slightly from frostbite. Anaktok, of course, was just fine.

Lawrence's report on the expedition concluded that navigation was the greatest obstacle to arctic flying. Magnetic compasses were *'very erratic at any time'* and in some localities *'it was utterly impossible and unsafe to attempt to navigate by them.'* Of the four types of compasses employed — earth inductor, magnetic, sun and aperiodic — only the last proved satisfactory; when used in conjunction with the turn and bank indicator it *'could be relied on with safety.'* As for shipping lanes, existing charts of the strait proved so inaccurate that Lawrence felt they would be *'a large and dangerous factor in the development of the strait as a commercial water route,'* and recommended an aerial photo-survey be made in order to prepare new ones.[7]

* * *

The aircraft handed over to Canada at the end of the war were now wearing out and new machines had to be purchased, all British or American designs modified to meet Canadian requirements and most of them built in Canada. Through the later 1920s several types of Vickers flying-boats were taken on strength, and DeHavilland Moths, Fairchild 71Bs and Bellanca CM-300 Pacemakers superseded the Avro 504s, DH4s and 9s. All three replacement types could be flown with either wheels or floats and the Fairchilds and Bellancas were good load carriers, excellent for bush flying. The Fairchilds would eventually be fitted with light bomb racks under the fuselage for practice bombing in army co-operation exercises.

In 1927 the RCAF felt able to afford two purely military purchases — nine Armstrong-Whitworth Siskin IIIs, then the RAF's first-line fighter, and six Atlases from the same British stable, the latest thing in army co-operation machines. Both types were biplanes. The Siskins, particularly, were what RCAF pilots dreamed about, capable of reaching speeds of nearly 290 kph, and the opportunity they offered to indulge in a combination of aerobatics and formation flying was one not to be missed. A Siskin Exhibition Flight was quickly formed and rapidly became a popular attraction at shows across the country.

One of the chosen few who flew in the Exhibition Flight (the precursors of today's

Vickers Vedette over Lac St. Joseph, Ontario. (CFP, PMR 98-164)

Oblique aerial photography from the front cockpit of a Vickers Vedette, 19 June 1931. (NAC, PA 68295)

Fairchild FC2W piloted by Flying Officer C.R. Dunlap, Port Alberni, British Columbia, 1931. Dunlap became an Air Marshal in the postwar RCAF. (CFP, PMR 98-107)

Snowbirds), formed and initially led by British exchange officer Flight Lieutenant Victor Beamish,* was Pilot Officer Fowler Gobeil.

I recall one day (in a three-plane training flight, flying in the left wing position) watching the upper man in a left-hand turn slide slowly but inexorably down the hill into Francis Victor's plane, his prop[eller] neatly chopping pieces out of Victor's right aileron. With our open cockpits, the noise was quite audible, and terrific!

Sitting about three feet off Victor's left wing, all I wanted to do was get the hell out of it, but I was too damned scared to break off. Not on account of a potential disaster, but because Francis Victor had impressed on us the cardinal sin of EVER breaking formation without the leader's signal. Sitting there in the choice grandstand seat, I watched in utter fascination as the unflappable Francis Victor gently waved his finger at the other pilot who was using his right aileron for a light snack. With his hand leisurely motioning at the end of a limp wrist, Francis Victor gently eased his boy away.... After we landed, all he said was a mild: 'My, chaps, we mustn't have any more of that. Let's go over it again.' And that was that.[8]

Flying at relatively low speeds compared with modern exponents of formation aerobatics, the Siskin pilots engaged in manoeuvres quite beyond the capacity of their successors.

*Two days before the CNE, Dave** decided that as the US Air Corps was bringing a full squadron of Curtiss Hawks, we had to do something spectacular, so he added to our few manoeuvres spinning in formation. We had one practice at Camp Borden consisting of No 1 and No 2 wing men moving out two spans from the leader, the three fliers then picked a point on the horizon,* [and] *put the aircraft in a right-hand spin, coming out on the picked point on the third turn. It worked!*[9]

In 1929 Parliament appropriated almost $6 million for government air operations — $4 million to be used for purely civil purposes, most of which in practice went to the RCAF, and nearly $2 million for flying training and military expenses. Left-wing pacifist MPs wanted civil and military budgets totally separated so that the RCAF vote could more easily be reduced or eliminated, but minister of national defence J.A. Ralston pointed out that both civil and military pilots were being trained at Camp Borden and that the station was being used 'principally as a training ground for civilian work.'

The RCAF vote, Ralston argued, was 'essential for the benefit of civil government operations and civil aviation.'[10] The only exclusively military equipment were the Siskin fighters, and no more were being acquired, as the policy now was to buy 'Moths and planes of that kind for the purpose of civil operations and to teach men in such operations.... We figured that we should turn our attention for the present entirely to civil government operations and civil aviation.' His point had already been driven home by the death of a civilian pilot in the crash of an RCAF Moth in April while the victim was training to be a flying club instructor.

The next year — 1930 — saw yet another increase in appropriations with the allocation of $7.5 million for air operations — fractionally more than thirty per cent of total defence expenditure, as opposed to the ten per cent of 1924. The RCAF's specific share was $2.5 million, and there were now 177 officers and 729 airmen on strength — but things were about to change for the worse. In October 1929 a stock market crash in the United States had ushered in the Great Depression, and in Canada the Depression led to drastic cuts in every aspect of spending. In 1931 the estimate for the air services was slashed by more than $2 million. RCAF expenditures were reduced by releasing twenty-eight other ranks, cutting back on forestry patrols and discontinuing the purchase of new aircraft.

Some years earlier work had begun on an new air station on the shore of Lake Ontario where, unlike at Camp Borden, there would be facilities for both land- and water-borne aircraft. Despite the financial squeeze, a start was made at transferring some of the Borden activities to Trenton, and gradually, over the following four

---

* First as a wing commander, then as a group captain, Beamish would become one of the RAF's outstanding fighter leaders during the Second World War.

** Flight Lieutenant Dave Harding, the officer who succeeded Beamish in command of the flight.

Vickers Vancouver at Havre St. Pierre, Quebec, Imperial airmail flights, 1932. (CFP, RE 19223)

Pilot Officer R.J. Clermont and Corporal F. Birch ready for gunnery practice, Camp Shilo, July 1934. The aircraft is an Avro 626 of No 112 (Army Co-operation) Squadron. (CFP, PMR 76-588)

No 6 General Purpose Detachment, Sydney, Nova Scotia, with Bellanca CH-300 Pacemakers for photography and transport. (CFP, PMR 98-165)

years, nearly all the training activities at Borden would be moved to Trenton, which would soon become the RCAF's major base.

The biggest cut came in 1932 when the air appropriation was slashed to the bone, leaving only $1.75 million to cover all expenses — a drop of $5.75 million from the vote of two years earlier. The army and navy were not cut nearly as severely, either in total or as a percentage of their earlier budgets. Indeed, the militia lost only slightly more than twenty per cent of its 1930 appropriation, compared to the air force's chop of more than seventy-five per cent — perhaps a tribute to the parliamentary power of the 'militia colonels' and the voting rights of many thousands of part-time militiamen who found their militia pay a useful supplement to their civilian salaries, or, in some unfortunate cases, their only source of income in a depressed society that had not yet developed a social welfare net.

The RCAF was compelled to release seventy-eight officers and another hundred airmen. *Ab initio* training was eliminated and other training curtailed, while a number of stations and sub-stations were reduced to a 'care and maintenance' basis. Further construction at Trenton and the improvement of several civil airports were suspended and civil operations cut back from the 30,000 flying hours of the previous year to 3,500 hours. The policy of no new purchases remained in effect. Back in 1924, the theoretical organization of the service had included a Non-Permanent Active Air Force (NPAAF), but at that time nothing had been done to actually establish it. Now, in an attempt to minimize the effects of the cuts, a part-time air force was finally created and three army co-operation squadrons were formed, one in Toronto, one in Winnipeg and one in Vancouver. Ground training commenced in the spring of 1933, but another year would elapse before any flying training was begun.

The major new activity in 1932 was one of co-operation with the Mounted Police in a campaign against rum-running, mostly off the east coast. These 'preventive' patrols accounted for almost half of all the civil flying carried out during the year. Prohibition had now ended in Canada (except in Prince Edward Island) and most of the liquor was destined for the United States where Prohibition was still the rule, but it was easier to land the stuff undetected on Nova Scotia's thinly populated coasts. Once ashore and broken up into smallish consignments, shiploads could be transported into Maine, New Hampshire or Vermont over little-used and unguarded byways by car or truck. Some liquor remained in Canada, since it was well worthwhile to 'import' it without the payment of excise taxes. Business was booming in the early 'thirties.

The easiest way to block this illicit trade was by intercepting the carriers within Canadian waters, or where cargoes were landed. Rum-runners could most easily be located and identified from the air, and to do that crews of three — pilot, navigator and an RCMP officer — manned Fairchild 71s equipped with floats.

The rum-runners normally stayed well offshore, beyond our range during the day, and then started to come in during the late afternoon or early evening to make their landings in the hours of darkness and then head out to sea again before dawn. Most of our patrols, therefore, were made as late as possible in the day to try to pick up the rum-runners on their way in....

We made no night patrols.... We had to rely on visual detection.... In addition, there were no aids to navigation; we had no wireless receivers, so even if there had been radio ranges, beacons or other aids, they would have been useless to us.... We had a transmitter but no receiver....

...for special patrols well off the coast — say, to 150 miles offshore — two aircraft flying in sight of one another had to be used and one had to carry a navigator. To call what we did navigation is a bit of a presumption. I had a small plywood table installed in the aircraft to which I could pin my charts; I had a pencil, a protractor and compasses to measure distances. After take-off we would fly over some easily recognizable point, such as the Halifax Citadel, on a course and airspeed that, with no wind, would take us over another point, such as Cape Sambro, so many minutes later. When the time was up we would pinpoint our actual position, and could then calculate the wind speed and direction. From then on everything was dead-reckoning. Seldom, if ever, was the wind over Halifax the

same as it was 150 miles off shore. All we could do was to watch the sea state as we flew along, and from it estimate, or guesstimate, wind changes.[11]

Throughout the 'twenties and 'thirties there was a steady drain of aircrew killed in training crashes, but one is always surprised at how few were lost on such hazardous operations as that described above.

Air mail was among those services curtailed by the onset of the Depression. However, an Imperial Economic Conference was scheduled to meet in Ottawa for two months during the summer of 1932, and there was a demand for letters to be moved between London and Ottawa as quickly as possible. As is so often the case, what politicians wanted, politicians got — at the taxpayer's expense. Arrangements were made for liners carrying the Royal Mail (as it then was, in both Britain and Canada) to be met by a Bellanca floatplane at Red Bay, on the Labrador coast, in order to fly the mail to Havre St. Pierre, Quebec. From there it was taken to Rimouski by Vancouver flying-boat and transferred to a wheeled Fairchild for delivery to Ottawa some twenty-four hours sooner than it would have got there by ship and train. The same procedure was followed in reverse for outgoing mail. It was a complicated and no doubt expensive scheme, but it worked.

* * *

By 1934 the government was beginning to see a need for an increased emphasis on military aviation, for Canada was still psychologically tied to Britain, and the British (and most of Europe) were now worrying over developments in Germany. In March, the day after MP Ian Mackenzie (a future minister of national defence) told the House of Commons that 'if it is necessary to carry out a real defence policy for Canada in the future, I think that the air force will play a most prominent part in that policy,'[12] the new German chancellor, Adolf Hitler, denounced the compulsory German disarmament clauses of the Versailles treaty which had ended the First World War.

In the subsequent debate over the air estimates, the current minister of defence, Donald Sutherland, observed that 'there was not really enough money in the last vote to get the full benefit of the air force as we had it,' there was simply not enough money allotted for flying and the airmen 'were grounded too much.'[13] The Depression was still playing havoc with the economy and cruelly punishing many innocent Canadians, but even so the Commons responded with an increase of $400,000 in the RCAF appropriation.* At the same time, the RCAF was able to acquire, at a reduced price, ten more well-worn Atlases, which needed to be reconditioned before they could be used for army co-operation training.

By the end of 1934 the Permanent Force component mustered 118 officers and 676 airmen, and the entire RCAF had 166 aircraft on its books, most of them suitable only for training or civil operations and all of them either obsolete or rapidly becoming so. There were twenty-eight military-type machines — fifteen Atlases, eight Siskins (one had 'died') and five Vickers Vancouver flying-boats, which were built to a civil design but had been converted for military use. For what it was worth, two further squadrons were authorized for the NPAAF, one bomber and one fighter unit, both in Montreal, and the part-timers ended the year with a strength of thirty-nine officers and 269 airmen.

Parliament increased the appropriations for all three services, with the air force again getting by far the biggest increase. That summer, Hitler announced a massive programme of re-armament that would give his Luftwaffe a first-line strength of some 2,300 aircraft by 1938. Britain planned to muster nearly the same number of machines by then, with (like the Germans) approximately twice as many bombers as fighters. Bombers with a transatlantic range were still in the indefinite future and Germany was not building aircraft carriers. The air threat to Canada was still indirect and remote, but, given Canadian ties to Britain, there was good reason to follow the European example and re-arm, and with more modern machines. Thus the RCAF increased its person-

---

* Appropriations for the militia and navy were both decreased again, however.

nel establishment by almost a third, and orders were placed for twenty-eight aircraft to replace those crashed or written off the previous year. The acquisitions would include six re-conditioned Westland Wapiti bombers and four brand-new Blackburn Shark torpedo-bombers.

Regrettably, the already-obsolescent Wapiti was no more than a modified First World War bomber, with open cockpits for the pilot and air gunner/bomb aimer. *'The aircraft was a beast,'* thought Flight Lieutenant C.R. Dunlap, on exchange duty with the RAF. *'It lumbered off the ground and struggled along with very unimpressive performance...it glided like a brick.'*[14] As for the Shark, torpedo-bombers made a lot of sense when the only feasible threat was one from seaborne raiders, but it too was a poor excuse for a contemporary aircraft. With a crew of two or three in an open cockpit, the Shark's maximum speed was only about 240 kph and its combat range no more than 400 kilometres. This at a time when the US Army Air Corps was already using — among other types — Martin B-10s and 12s with enclosed cockpits, 100-kph faster with a range more than double that of the Shark while carrying a greater bomb load.

Canada's defence problems were essentially the same as those of the United States, and it would have made more sense to switch to American designs, which were more appropriate to the North American strategic situation. Nevertheless:

> The Royal Canadian Air Force is organized on Royal Air Force practice, and therefore Royal Air Force equipment fits into the Canadian scheme better than American equipment.... We use the same establishment [as the RAF], and to get the best results should use the same equipment. If we change the equipment we must change the establishment.... Similarly, if we change the type of aircraft and go in for the American product, it means that we must also change our supply of bombs, guns, and instruments of all kinds, for the American aircraft are designed to take American accessories.... This is not an insuperable job and could be incorporated in the manufacture if we are building this type of aircraft in Canada.[15]

That was not a step that the Senior Air Officer was willing to recommend, however, and even if he had done so it seems unlikely that either his superior, the Chief of the General Staff, or the Cabinet would have acquiesced. The British connection was still too strong and, besides, British machines were generally cheaper.

RCAF appropriations rose again in 1936, as tensions in Europe continued to mount. In March, Hitler denounced the Locarno Pacts of 1925 and re-occupied the Rhineland; in July, the Spanish Civil War broke out. Canada's response was to order three more Sharks, five Supermarine Stranraer flying-boats for maritime reconnaissance and a number of new trainers. The Sharks and trainers would be delivered in 1937, but the first Stranraer would not go into squadron service until late in 1938, by which time it would be yet another obsolescent machine.

The first service type to be built in Canada since the First World War, the Stranraer was constructed in Montreal by Canadian Vickers to a British design. A twin-engined biplane with a crew of five, it had a range of 900 km at a speed of 175 kph and could carry 570 kgs of bombs or depth charges on racks under the lower wings.

Personnel strengths in both the Permanent and Non-Permanent components of the RCAF were increasing, but the number of aircraft was actually dropping. By March 1937 there would be only thirty service aircraft on charge, including the eight distinctly obsolete Siskins and fourteen more-or-less obsolete Atlases, together with forty-six trainers. The other eight machines were flying-boats. However, service flying training rose to a total of 11,000 hours during 1936 and civil operations except for photo-survey were finally transferred to a new Department of Transport, leaving DND responsible for military flying alone.

When Parliament assembled, in January 1937, its members were perhaps more concerned with national defence than they had ever been in the past. During 1936 both Germany and Japan had denounced international agreements that inhibited their military development and Germany had unilaterally re-occupied the

Siskin aerobatic team: Flying Officer E.A. McNab, Flight Lieutenant F.V. Beamish and Flying Officer E.A. McGowan. (CFP, PMR 82-162)

Flour-bomb dropping as part of army co-operation exercises, 1934. (CFP, PMR 98-168)

Blackburn Shark drops a torpedo. (NAC, PA 141380)

Rhineland. In October, the creation of a Berlin-Rome 'Axis' strengthened the positions of Germany and Italy in relation to Britain and France.

The Japanese, who had already taken Manchuria, were threatening China (they would invade in July) and were beginning to talk in terms of a Greater East Asia Co-prosperity Sphere, a concept which, if it came to fruition, would certainly imperil British colonies in the Far East. They did have aircraft carriers, and were building more. There was no real likelihood of invasion if a general war should break out, but whoever the enemy might be, Canada was likely to become involved on the British side. Fast sea-raiders, of the heavy cruiser or pocket battleship class, might then bombard ports on either coast or even put raiding parties ashore, while carrier-borne aircraft could certainly attack west coast ports.

Consequently, the RCAF appropriation rose to $11.75 million, an increase of almost $5 million over the previous year, and an increase in strength of more than fifty per cent was authorized for the Permanent Force. The NPAAF would be almost doubled in size. However, most of the money would go to the purchase of additional aircraft. Among them would be twenty-four new bombers, eighteen coastal reconnaissance machines, eleven more torpedo-bombers, seven flying-boats and twelve two-seater fighters.

These last, in the form of single-engined Fairey Battle low-winged monoplanes, designed as a light bomber, would prove a serious mistake, for the Battle was a grossly under-powered machine, totally unfit for combat. It *looked* a likely candidate, and the Canadians, with their airfields relatively far apart and a need for long-range fighters, apparently thought it could be used in the latter role despite its poor armament of one fixed, forward-firing machine-gun and one flexibly-mounted gun in the rear cockpit. Another procurement mistake.

* * *

In so many respects the RCAF was technologically a generation behind. Air Commodore Len Birchall's reminiscence of a 1938 'mercy flight' by a Fairchild to bring a seriously ill woman to hospital from Sable Island — little more than a sand bar some 160 kilometres off the Nova Scotia coast — gives the flavour of the times.

We had very carefully calculated just how far we could fly out to sea before we had to turn and come back — there was no fuel out there anywhere. The only additional navigation aid we had was a sheet of clear plastic set in the floor with various lines inscribed on it. To use this device you sat and gazed down through it at the little bits of froth on the wave tops and calculated your drift from your changing relationship with them. Then you had the pilot kick 60 degrees off track while you took another drift reading. Then 120 degrees back you came for a third reading, after which you would straighten out to what you thought had been your original track. Now you had three drift readings and from these you calculated your wind vector, and away you went. When I remember that trip to Sable Island — I would no more try that trip today than.... Yet it was just routine stuff, almost, in those days.[16]

As world tensions heated up it was becoming apparent that the RCAF's condition was pitiable, and the Munich crisis of September 1938 brought that reality home to the Cabinet. Canadian air defence was something of a farce. The planned east coast coastal artillery co-operation squadron had still not been formed, and now, when the need arose, No 2 (Army Co-operation) Squadron had to shepherd its time-worn Atlases from Trenton to Halifax in order to fill the gap. It took them an inordinately long time, and 'With them went all the other aircraft the RCAF was able to muster: a total of thirty-nine, among them six Sharks, the only remotely modern military types. Thirteen others were obsolete service machines, the rest were civil aeroplanes. Only twelve of the thirty-nine could carry effective bomb loads.'[17] That hardly mattered: there were no bombs weighing more than twelve kilograms.

This dose of harsh reality led the Cabinet to despatch a mission to Washington 'to buy as many aircraft...as $5 million would purchase.' That sum would have paid for eighteen Douglas 18A bombers, fifteen North American 25 reconnaissance/co-operation machines and fifteen Seversky P 35 fighters, all not only reasonably modern machines but also well suited to Canadian needs. However, a settlement of the crisis by British prime minister Neville Chamberlain — 'peace in our time,' he proclaimed on returning from Munich — led Ottawa to have second thoughts. Before a deal could be completed, the financing was cancelled.

Nevertheless, the increasing importance of the air arm was formally recognized — since there was no great expense involved! — by its final separation from the Militia. Until November 1938 the RCAF had come under the ultimate authority of the Chief of the General Staff; but that subordinate status came to an end on the 19th, when it was decided that 'the control and administration of the Royal Canadian Air Force will be exercised and carried out by the Senior Air Officer who will, in this respect, be directly responsible to the Minister of National Defence.'[18] A month later, in keeping with this new independence, the title of Senior Air Officer was changed to that of Chief of the Air Staff, the first holder of the appointment being the former Senior Air Officer, Air Vice-Marshal G.M. Croil.

Internationally, the threat was still there, though not quite as imminent as it had appeared a year earlier. When Parliament reconvened in January 1939 the minister of national defence proposed that the unprecedented sum of $60 million be appropriated for defence spending, of which the RCAF's share would be $23.5 million. Long-range planning called for an eventual establishment of 525 officers and 4,500 men in the Permanent Force and 220 officers and 2,000 men in the Auxiliary air force. Even this year, the strength of the former was to be raised to 273 officers and 2,172 men. By the spring of 1940 the service expected to have 250 aircraft on strength and during the current year orders would be placed for another 107 machines. They would include more single-engined fighters and army co-operation aircraft, twin-engined bombers, twin-engined general reconnaissance machines and flying-boats. Lamentably, all would be British designs, although many would be built in Canada.

The intention was to build up operational strength to eleven permanent and twelve auxiliary squadrons. Eight of them would be assigned to the defence of the west coast, nine to the east coast, with the remaining six in central Canada 'for use as a reserve in accordance with exigencies that might arise.'[19] No mention was made of the probability of support for a Canadian expeditionary force, or the possibility of reinforcing the RAF in any way.

There was every reason to augment and modernize the air arm. At the end of March — when war was only five months away, although no one could know that — the RCAF had only twenty so-called 'first-line' machines on strength, ten of them the bungled Battles, although ten Hawker Hurricanes — at last, a really up-to-date fighter! — would be added to the roster within days. There were another twenty-two obsolescent service types, and twelve civilian-pattern aircraft were being converted to military standards. The rest were either elementary trainers or entirely obsolete. Those figures made nonsense of the minister's claim that 'the first line of defence for the Dominion of Canada must be the air force.'[20]

The Hurricanes — Canada's first truly contemporary military aircraft — were as dramatic an acquisition as the Siskins had been twelve years earlier. They arrived at Vancouver aboard a freighter in February, and first had to be uncrated and assembled. Low-winged, single-seater monoplanes, with eight .303 machine-guns mounted in the wings and retractable undercarriages, they could fly at 520 kph and reach a height of 10,000 metres. However, since they had been designed for the RAF's defence of the United Kingdom, their range was very limited.

'Ten little fighter planes, sitting on the tarmac....' Very soon there were only nine, when a pilot making his first flight in one swerved while trying to take off and hit another aircraft. The Hurricane was a wreck although the pilot was unhurt. 'Nine little fighter planes, flying off to Calgary....' Suddenly there were only eight, when one crashed near Mission, BC, and the pilot was killed.*

---

* Another shipment of eleven Hurricanes arrived on the eve of war.

The international situation was steadily deteriorating, with Hitler's threats increasing in intensity and frequency. Soviet Russia and Nazi Germany signed a surprising non-aggression pact on 21 August that left Hitler free to move against Poland without fear of Russian retribution, and a week later the RCAF's eight Permanent Force squadrons were en route to their war stations. On 1 September 1939 Hitler attacked Poland, and on the 4rd, the day after Britain and France declared war, all of the RCAF was placed on active service while the nation waited for Parliament to re-assemble and decide on Canada's role.

That decision was a foregone conclusion, and on the 10th Canada declared war: the RCAF was about to meet its first — and so far, only — major test. Fortunately for the service, and for Canada, it would be a long, drawn-out conflict. Before the German and Japanese surrenders there would be time for the RCAF to develop into a significant protagonist in the air war.

Chapter III

TRAINING FOR WAR

Armchair air marshals like to focus on the more intriguing perspectives of war — the challenges and hazards of operational planning, the benefits of superior technology and the impact of small groups, or even individuals, on the ebb and flow of battle. However, professional airmen (and soldiers and sailors, too) appreciate that another, more remote, ingredient, logistics — which is simply the acquisition and supply of men and materiel — is every bit as important to success, if not quite as exciting to read about.

Without sound and solid logistics behind them, in the long term there can be no successful strategy, operations or tactics. That was never more true than in the aerial undertakings of the Second World War, which was won more by brute force than by any subtleties of campaigning or superior technology. The creation of that force was a matter of good logistics, and it demanded long-term planning and endless effort. Many thousands of aircraft were needed, and hundreds of thousands of aircrew. But aircrew, especially the key category of well-trained, competent pilots, could not be churned out in a matter of weeks, or even months. Supplying and training aircrew in large numbers was arguably Canada's greatest single contribution to victory.

American president Franklin D. Roosevelt once described Canada as 'the aerodrome of democracy' — although those were not actually his words! Lester Pearson, a future prime minister of Canada, was then first counsellor at the Canadian embassy in Washington.

> Once in Washington I was even a ghost writer for President Roosevelt, though he may never have known it. The President wished to send a message of congratulation to Mr. [Mackenzie] King on the third anniversary of the British Commonwealth Air Training Plan, a project in which Canada now took a great and justifiable pride. I was surprised when a friend on the White House staff, ignoring all rules of diplomatic propriety and without telling the State Department

anything, asked me whether I would be kind enough to do a draft of the message for the President. So on 1 January 1943 the Prime Minister of Canada received a very impressive letter lauding Canada as the 'aerodrome of democracy' drafted by me but signed by the President of the United States.[1]

The phrase was a memorable one, but even metaphorically it was far from true. A more precise description might have been 'the aerodrome of the British Commonwealth,' for when the Americans were brought into the picture Canada came a poor second in production of aircrew. Although the United States did not officially enter the war until the Japanese attack on Pearl Harbor, on 7 December 1941, the US Army Air Forces *alone** graduated (in round numbers) 193,000 pilots between July 1939 and August 1945, together with 50,000 navigators and 45,000 bombardiers. The British Commonwealth Air Training Plan (BCATP for short, but the British themselves — and many anglophiles — were inclined to call it the Empire Air Training Scheme, or EATS) produced no more than 132,000 aircrew of all categories, composed (again in round numbers) of nearly 73,000 Canadians, 42,000 from the RAF (including Free French, Norwegians, Poles, Dutchmen, Belgians and Czechs), 10,000 Australians and 7,000 New Zealanders.

* * *

As early as 1936, the British had sought to establish RAF flying training schools in Canada, as had been the case during the First World War. Such schools would train both British and Canadian candidates for the RAF. Britain was a small island, heavily populated in those parts suitable for airfields, subject to dubious, uncertain weather and, in the event of war, under the risk of enemy air attack. It was positively unsuitable for training large numbers of aircrew. Canada, on the other hand, was virtually immune to hostile threats, blessed with relatively predictable weather and possessed of vast expanses of land suitable for airfields; and, in the event of war, it was not as far away from Britain as Australia or southern Africa, the other suitable sites for large-scale training.

However, the British proposal proved unacceptable in Ottawa, where the Cabinet whispered its intention to 'establish training schools of its own' and feared 'the situation might give rise to competition between governments in the matter of fields, pilots, equipment and the like.'[2] Such talk came cheap. Two years later, the only service flying school in Canada was still the small one at Trenton that had been in existence for a decade.

In May 1938 the British tried again, this time endeavouring to appease Canadian sensibilities by suggesting that control and supervision of such RAF establishments might be left entirely in RCAF hands. From their perspective there was more than flying schools at stake. Behind their perfectly understandable enthusiasm for the schools was a hope that getting Ottawa involved in this particular scheme might help break down Canadian reluctance to get involved in other, more comprehensive, imperial designs.

Prime Minister Mackenzie King, on the other hand, while he had become more open to the idea of RAF training in Canada (on Canadian terms), was determined to evade broader commitments. Ten years later, towards the end of a long political life, King would note in his diary:

> I have been for some time the only Prime Minister [in the Commonwealth] who was Prime Minister prior to the war and who has continued as Prime Minister since.... This I put down to having taken just the opposite view to that taken by those of an imperialistic outlook.[3]

In 1938 the whole issue became unbelievably confused, with national interests entangled with political advantage and questions of who would pay for what. Negotiations teetered on for a year. In the end, however, King formally rejected the British proposal, while offering to train some candidates from the United Kingdom in RCAF schools. The first batch were scheduled to arrive in mid-September 1939. They never

---

* The US Army Air Corps became the Army Air Forces on 9 March 1942. The US Navy had its own air arm and trained its own aircrew. There would be no independent US Air Force until July 1947.

A DeHavilland DH.82 Tiger Moth. Canadian winters forced the adoption of a 'greenhouse' canopy. (CFP, PL 3582)

appeared, for the British went to war on 3 September and were too busy for the moment to concern themselves with such trivial matters.

Push having now come to shove, Canada very soon joined Britain and France in their war on Germany. Negotiations for a much larger training scheme were re-opened, and by the end of November a provisional Agreement called for airmen from Britain, Australia and New Zealand — together with Canadians — to be trained under Canadian auspices. It was agreed that Dominion candidates should be trained in numbers proportionate to populations — fifty-seven per cent Canadian, thirty-five per cent Australian and eight per cent from New Zealand. Only ten per cent would be British, for, despite the inherent difficulties recounted above, the RAF still intended to train most of its aircrew at home.

Calculating on a three-year Plan, the cost was estimated to be in the region of $600 million — a phenomenal amount for the time and several billions by today's standards. The United Kingdom anticipated contributing nearly $200 million, mostly in aircraft, spares and equipment. Canada would be responsible for eighty-one per cent of the balance, Australia for eleven per cent and New Zealand for eight per cent. Australia and New Zealand would carry out their own elementary flying training and their smaller shares, as well as their smaller numbers, reflected that.

The question of identifying graduates according to their nationality also arose and, although no one foresaw it at the time, this was to become a major embarrassment to the Canadians. Article XV of the Agreement in draft form stipulated that 'pupils of Canada, Australia and New Zealand shall, after training...be identified with their respective Dominions, either by...organizing Dominion units or in some other way.' Mackenzie King, however, wanted 'a clear and unequivocal statement that, at the request of the Canadian Government Canadian personnel from the training plan would be organized in RCAF units and formations in the field.'[4]

King pushed hard and won his point. Initially, however, this matter of national identities was complicated by the fact that all operating costs of Dominion squadrons serving overseas were to be borne by the British, supposedly to compensate the Dominions for their relatively large expenditures on the Plan; and that in the Canadian case most groundcrew would have to be from the RAF, since all those that the RCAF could produce in the foreseeable future would be needed to service the burgeoning number of aircraft in BCATP schools. The first half of that dilemma would, in January 1943, lead the government to take full financial responsibility for all its squadrons — something that it should have done from the beginning; the second would eventually solve itself as trade schools produced the necessary craftsmen.

King finally embraced the Plan for political as well as military reasons, viewing it as a project that would demonstrate Canada's commitment to victory, would not be too repugnant to Quebeckers and would steer him well clear of the rocks of conscription. At the same time, however, he needed the support of anglophiles in the rest of Canada, so he persuaded the British prime minister to declare that the Plan 'would provide for more effective assistance... than any other form of military co-operation which Canada can give.' But, Neville Chamberlain added in a sly attempt to undermine King's struggle to limit Canadian commitment, the British government 'would welcome no less heartily the presence of land forces in the theatre of war.'[5]

In such devious matters King had no equal. Merely adding a brief phrase of his own in announcing the birth of the Plan on 17 December, he told the nation (over the CBC's airwaves) that the British would welcome Canadian troops 'in the theatre of war' — and added his own phrase, 'at the earliest possible moment.' Even as he spoke, the 1st Canadian Division was en route to the United Kingdom and would obviously arrive 'at the earliest possible moment.' That met Chamberlain's shrewd thrust while also, King hoped, satisfying his anglophile constituency. There would be no need to send more men, no need to even think about conscription!

Planning to implement the Plan began almost immediately, but although work went on apace there seemed no great urgency. All was still quiet on the Western Front — until April 1940, when Hitler abruptly attacked Denmark and Norway. Denmark was taken in a day and, while British and French expeditionary forces struggled to extricate themselves from Norway, on 10 May the *Blitzkrieg* fell on Holland, Belgium and France. In six horrendous weeks all four countries fell under the Nazi jackboot, leaving Britain in apparently desperate plight.

Mackenzie King's attitude changed dramatically, as did indeed most of Canada's, in the face of Hitler's successes. Until that time, only one BCATP school* — the Central Flying School at Trenton, already absorbed into the Plan, and devoted to spawning instructors — had begun work. Now there was every reason for haste and, while planning continued, other schools quickly came on stream.

Each Dominion was responsible for its own elementary pilot training in accordance with the BCATP Agreement. In Canada's case that obligation was laid upon the country's twenty-two civilian flying clubs, which worked on a cost plus five per cent basis. The first four elementary training schools (EFTSs) opened on 24 June, three days after France capitulated, and by the end of July the number had doubled. Before the end of the year the number would double again, and by September 1941 every flying club in the country would be running one.

The first of ten civilian-operated air observer schools (AOSs), most of them sharing facilities with EFTSs, was set up, and the ten would subsequently be supplemented with six Air Navigation Schools. Most navigation training was more conveniently conducted in classrooms and the relatively few flights needed for airborne training could be flown from airfields kept busy by primary trainers.

Ultimately there would also be eleven Bombing and gunnery schools, four wireless schools and (from July 1944) a flight engineers' school. Flight engineers, who replaced second pilots in heavy bombers during the summer of 1942, were expected to keep track of fuel consumption and monitor the engines, making what adjustments they could in flight to maximize performance. They were also given the minimal amount of pilot training that would enable them to land an aircraft should the pilot be killed or incapacitated! (The BCATP was very late in getting into that particular field; consequently, most flight engineers, even in Bomber Command's Canadian group, were British, trained in the United Kingdom by the RAF.)

There was much competition, especially among prairie communities, to get a flying school established in their vicinity, since the schools brought significant economic benefits to neighbourhoods devastated by the Great Depression. Even the railways got into the act:

> At a Souris Board of Trade committee meeting a few weeks ago, *The Plaindealer* drew the attention of the members to the fact that up to the present every air school established in Manitoba has been located on the Canadian National Railway. As Sir Edward Beatty [chairman of the CPR] says, the railway employees of the Canadian Pacific have an equal right to share in the stimulated business created by these activities of the nation.[6]

Whatever the justification for such claims, there was little doubt about the political influences brought to bear in establishing schools. Most Liberal ridings (including the prime minister's constituency of Prince Albert) acquired one or more, and they were usually first in line, with CCF ridings that had previously been Liberal not far behind. However, much to Mackenzie King's chagrin, Davidson, Saskatchewan, which lay in the constituency of a new Conservative MP named John George Diefenbaker, did get one of the last EFTSs to be set up.

Meanwhile, the shock of *Blitzkrieg* had weakened Mackenzie King's resistance to the idea of RAF schools. In July 1940 the British had asked if they could move four RAF Service Flying Training Schools (SFTSs) to Canada, for after the fall of France operational pressures on British airfields left little room for advanced training. Not only were they now acceptable, replied Ottawa, but if the British wished to

---

* Under the aegis of the RCAF, eight civilian flying clubs had been providing elementary instruction for RCAF candidates since June 1939 and graduates had been receiving service flying training at Camp Borden since November.

transfer still more, then space for them could also be found. Promptly, the RAF responded with four more SFTSs, two air observer schools, a general reconnaissance school, a navigation school, a bombing and gunnery school and a torpedo-bombing school. Eventually, the RAF operated twenty-six schools of one kind or another under the aegis of the BCATP.

The last stop for BCATP graduates — or at least those of them destined for Home Defence squadrons — was an Operational Training Unit (OTU) where aircrew were familiarized with the latest developments in the field (or perhaps sky?). They flew the same types of aircraft that they could expect to use on operations, although the machines they used were often tired old ones that had been replaced on operations by newer variants. Moreover, all OTUs in Canada lacked some equipment deemed essential for proper operational training, such as VHF radios and beam approach landing systems, that were simply not available in Canada.

The first of four RAF OTUs, at Debert, Nova Scotia, trained Lockheed Hudson crews destined for Coastal Command. Despite the inevitable holes in their operational training, graduates subsequently transferred themselves to a theatre of war by ferrying new machines, produced in the United States, across the Atlantic.

The first of three RCAF OTUs opened at Bagotville, Quebec, in July 1942, using ancient Hawker Hurricanes to train fighter pilots for Home Defence squadrons. (Fighter pilots destined for Europe and heavy bomber aircrews usually received their operational training in Britain.) The two others were at Patricia Bay, on Vancouver Island — a site shared with an RAF OTU, where both trained maritime reconnaissance crews — and briefly at Boundary Bay, near Vancouver, before moving to Abbotsford, British Columbia, where it trained heavy bomber crews mostly destined for the Far East theatre from April 1944.

* * *

At the end of April 1940 the first 164 Canadian BCATP candidates were posted to No 1 Initial Training School for a four-week course that included lectures in mathematics, basic navigation, aerodynamics and armaments. Six of the chosen few failed this first stage of training or dropped out; ninety-two were selected for pilot training, forty-one for tutoring as observers* and twenty-five for wireless operator/air gunner instruction. These were probably fairly typical proportions for subsequent intakes.

Those assigned to pilot training were then posted to an EFTS (where they were joined by a number of British candidates) and taught the fundamentals of flying in Tiger Moths or Fleet Finches — both two-seater, open-cockpit biplanes, although a perspex hood was fitted during the winter. Given the civilian nature of the EFTSs, most instructors were civilians. Whether servicemen or civilians, their first task was to introduce their pupils to some of the less gratifying possibilities of flight.

A piece of paper floated idly in front of my face, went up and settled on the coupe top, then slipped back to the floor where it belonged. I looked out and the ground loomed closer. Then all was serene. All, that is, except the waves of nausea creeping from beneath my belt to the vicinity of my throat. 'The roll off the top. A combination of the last two manoeuvres,' announced Mr. Potts calmly.... The wave in my stomach lurched, surged forth, and hung in mid-air, along with the piece of paper.[7]

Some pupils were never airsick. Some overcame it, while others never did and were soon 'washed out.' Yet others simply could not master the elementary intricacies of flying. What happened when a pupil, flying solo, tried to land 'about thirty feet above the ground'?

The Finch hit on all three wheels.... The fuselage bent in the middle, the wings sagged, the [landing] *gear crumpled. Out of the cloud of dust the Fleet staggered on flat tires, wires trailing under the broken tail, smashed prop beating the ground. The student taxied to the line, swung around and parked neatly between two intact trainers, stiff upper lip to the end. He became a navigator....*[8]

---

* Not until 1942 was the 'observer' trade, which embraced navigation and bomb aiming, split into the specialized 'navigator' and 'bomb aimer' trades.

Fleet Finch in the mud at No 9 Elementary Flying Training School, St. Catharines, Ontario, March 1941. (CFP via F. Patterson, PMR 75-352)

Hurricane XIIs of No 1 Operational Training Unit, Bagotville, 1943. (CFP, PMR 76-268)

Piggyback Anson crash, Macleod, Alberta, 1941. (CFP, PL 988)

For those whose basic skills proved adequate, the next step was an SFTS run by the RCAF. There they were joined by EFTS graduates from Australia and New Zealand, together with more students from the United Kingdom. The majority of the twenty-nine SFTSs were settled on the Prairies — in Manitoba, Saskatchewan and Alberta. Half a dozen were to be found in Ontario. Two, originally ensconced in New Brunswick and Quebec, were subsequently moved to Saskatchewan.

Whether it was to be a single-engine school devoted to potential fighter aces or a twin-engine school for bomber, general reconnaissance or transport pilots depended upon the assessments that had been made at EFTS.

Those posted to single-engine schools found themselves flying North American Harvards, two-seater low-wing monoplanes with a single hood over both pilots. Their introduction to the much more complicated, more powerful Harvard brought a degree of bewilderment as great as that initially experienced with the Tiger Moth.

Climbing up on the left wing, we peered into the open front cockpit and caught our breath. The wide spaces on each side of the aluminium seat were crammed with handles, wheels and levers of all shapes, sizes and mysterious uses. The broad instrument panel contained a hopeless confusion of black-faced dials and toggle switches. More handles protruded from beneath the instruments and between the big rudder pedals.[9]

Harvard pilots learned basic aerobatics, formation and instrument flying, and how to recover from stalls and spins.

After five or six spins, but no lower than three thousand feet [900 metres] *the stick was pushed fully forward and opposite rudder applied. This increased the speed of the spin which seemed to make it worse.... It was the tightening of the spin and the higher speed with a steeper angle of descent that always worried me.... Remarkably, the Harvard snapped out of the spin very suddenly and it was a simple matter to pull back on the stick and apply reverse rudder to be flying straight and level again.*[10]

Those who went to twin-engine schools trained on Avro Ansons or Cessna Cranes, monoplanes with cabin accommodation for three or four. Reflecting their probable futures, Anson or Crane students spent more time on cross-country navigation, formation and instrument flying. Formation flying could involve more risks than aerobatics.

I had arranged for two other aircraft to formate...with our own Anson.... While we were getting into position, a sudden movement to our right and behind us caught my eye. I looked again, and saw an Anson coming up on us, moving far too quickly. I told the trainee who was flying, as calmly and quietly as I could, 'Okay, I'll take her,' and grabbed the controls...just seconds before there was a tremendous crash and the Anson rocked and staggered in the air. The other aircraft had smashed into us just behind the main cabin; its wing struck our fuselage and, as I glanced quickly out of the window, I saw the stricken aircraft, with one wing torn away, flip over and plunge towards the earth....

The controls were sloppy under my hands — it was a bit like trying to fly through mud — but they did respond, and I was able to get the aircraft turned around and slowed down and we limped back to Hagersville airport.... The two students in the other aircraft were killed, and the rest of us survived mainly by a fluke. ...the original design had a gun turret just behind the cabin. In the training aircraft, the weight represented by this gun turret was replaced by a large block of concrete, and it was into this concrete that the out-of-control aircraft had crashed. A few feet forward and none of us would have survived.[11]

Given the circumstances under which the Plan was put into effect the casualty rate was surprisingly low for most of the war. In 1940–41 over 11,000 training hours were flown per fatal casualty, and in 1941–42 results were even better — 14,000 hours per casualty. In 1943–44 the rate rose to one in more than 20,000 hours. Only in 1942–43 was there a dramatic 'blip' in the figures, with one fatal casualty for each 1,725 hours flown, for reasons that remain unclear. Meanwhile:

We knew that over 25 per cent of all pilots were being washed out.... We just had to get our pilots' Wings. Still more important, we just had to get

An instructor briefs his pupil before the latter goes solo, No 2 Service Flying Training School, Uplands. (CFP, PL 1137)

Avro Ansons from No 7 Service Flying Training School, Macleod, Alberta. (CFP, PMR 74-257)

Aerial gunnery training in Fairey Battles, Jarvis, Ontario. (CFP)

overseas. Damn those ground school exams and damn that miserable flying instructor who couldn't see how good a pilot you were. Our motivation came from within. No one needed to coerce us. There were no pep talks on how badly pilots were needed to fight the war. They weren't necessary. We read the papers. We knew Hitler was stomping through country after country; was rounding up Jews. We knew that Buzz Beurling, the kid from Verdun, Quebec, was shooting Germans and Italians out of the sky over Malta, his Spitfire a formidable threat to enemy formations. God, how great it all seemed...if you could only get your Wings.[12]

BCATP training was carried out without electronic navigation aids, although by 1942 such devices were being fitted on operational aircraft.* At air observer schools, the candidates' task was to master the art of navigation by 'dead reckoning' — a combination of observation of ground features, careful measurement of varying airspeeds and compass bearings, and estimates of wind speeds and direction. The results might be checked against sun or star 'fixes' whenever either could be seen, but taking an accurate sight with a sextant was as much art as science and required a great deal of practice — much more than BCATP navigators could expect to get.

Moreover, there was no industrial smog and daytime visibility was usually excellent, while at night there was no black-out and lights on the ground could easily be correlated with towns and villages on the map. Training under these conditions would prove to be of limited value when the airmen needed to find their position over Germany. No wonder that in mid-1941 only a minority of bomber crews could find their way to within five miles of their target over the darkened, smog-ridden Ruhr!

Until the summer of 1942 observers were both bomb aimers and navigators. It was too much for one man. Air Chief Marshal Arthur Harris, the newly appointed commander-in-chief of Bomber Command, made the point clear. Because bombers could not survive over Germany by day without fighter escorts, and the British had no long-range fighters, nearly all such bombing was done by night. Consequently:

> There was an obvious need to introduce the air bomber.... The navigator had more than enough to do...to get the aircraft within a few miles of the target, especially when making the run-up.... Apart from all the other difficulties...the work he had done [over a carefully illuminated chart table] as a navigator left him no time to get his eyes conditioned to the darkness, which he would have to do before trying to spot the aiming point.[13]

In June 1942 the observer trade was phased out, to be replaced by specialist navigators and air bombers. Two new kinds of navigator also appeared, in the categories of navigators 'B' for aircraft such as medium bombers and maritime reconnaissance aircraft that lacked the sophisticated electronic aids now being installed in heavy bombers, and navigators 'W', with wireless training, for twin-engined night fighters, intruders or light bombers in which there was no room for a wireless operator.

The original BCATP Agreement was due to expire in March 1943, but the war was still in full swing. A revised Agreement was negotiated, extending the Plan for another three years, and this new Plan officially incorporated the RAF schools and added several more RCAF ones, enlarging some of those that already existed. 'The emphasis on training swung from quantity...to quality. Yet quantity remained an important factor.... In December 1942 the monthly output, the greatest to date, rose to 4,332.'[14]

There was only one major source of disagreement over the revised Plan, and that related to commissioning policy. Prior to the new Agreement, fifty per cent of pilots and observers and twenty per cent of wireless operators/air gunners were to be commissioned, half on graduation and the other half later, based on performance in the course of operations. Air Minister C.G. ('Chubby') Power disliked that quota system, believing that it was based on the British idea that only members of a certain social class were suitable to hold commis-

---

* Even the best of them were not comparable with the sophisticated equipment of today, however.

Leading Aircraftman Jean Paul Sabourin of St. Isidore-de-Prescott, Ontario, under pilot training, No 1 Elementary Flying Training School, 1940. (CFP, RE 67-2118)

Pilot Officer Sabourin shortly before being posted overseas. (CFP, RE 67-2119)

The grave of Flight Lieutenant Sabourin, DFC, killed in action, North Africa, 16 September 1942. (CFP, RE 67-2117)

sions.*[15] He, on the other hand, argued for 'the absolute justice of making every member of aircrew an officer.'

Douglas Harvey, then a flight sergeant, bomber pilot, has made his case effectively:

The three officers in my crew lived in permanent barracks.... They had separate bedrooms, a kitchen, laundry, and batmen to look after their clothes, polish their shoes, make their beds, and serve up tea or whatever....

The sergeants lived off the base, five miles away.... Our 'bedroom' consisted of a room fifty feet long.... A row of iron cots stretched along each wall.... No cupboards, no dressers, not even a book on the wall. You kept your clothes in a duffel bag beside your bed....

The longer I spent in Benningborough Hall, sharing a bathroom with twenty-five others and living without the least amenities, the more bitter I became towards a system that rewarded some more than others for the same duty.... Only a few hours before I had led my crew, giving orders and demanding instant obedience.... But as soon as I landed I reverted to a subservient role. I couldn't understand such a stupid system.[16]

'Absolute justice' was an unlikely conclusion, however. Trying for a compromise, Canada argued that all pilots and observers should be commissioned on graduation, with twenty-five per cent of the other categories of aircrew getting commissioned then and another twenty-five per cent later. The British would not accept that, and finally the two sides simply agreed to differ. The RAF adhered to the old, established policy. The RCAF now chose to commission all pilots, observers, navigators and air bombers, but, for the time being, stuck with the existing quota for wireless operators and air gunners.

Under the revised Plan, Canada was to supply approximately half of the annual intake of candidates while the British would provide not less than forty per cent. The balance would come from Australia and New Zealand. There was some concern on the RCAF's part as to whether Canada would be able to meet its quota, however. At the beginning of April 1943 the pool of airmen awaiting training numbered nearly 10,000 but new recruits numbered only 1,300, and in May the intake dropped to 1,200. Planning called for 2,900 men to begin training each month, so that the reserve dropped significantly, to 6,000 by the end of May.

The RAF, which was by this time relying on the RCAF for a quarter of its aircrew, also fretted. The British chief of air staff, Sir Charles Portal, put his worries on paper.

> I am sure I need not emphasize the importance of maintaining, at this critical stage of the war, the output of aircrews adequate to meet the growing output of aircraft, and I should be much obliged if you would give the utmost consideration to the ways and means of maintaining the Canadian quota upon which, for quality as well as quantity, we so largely depend.[17]

Struggling to ensure its quota would be filled, Ottawa lowered the age limit, set at seventeen and a half in October 1941, by six months. However, boys enlisted at that age were not eligible to actually commence flying training until they reached their next birthday; and since it took, at a minimum, a year for them to work their way through the training system, no one would graduate before their nineteenth birthday and most would be at least nineteen and a half before they joined a squadron.

Another untapped source was found in the ranks of ground tradesmen. A Women's Division of the RCAF had been formed in July 1941, and now a spirited recruiting campaign emphasized that 'They Serve That Men May Fly.' Their numbers promptly doubled to 15,000. Some were trained as instrument mechanics, a trade at which their nimble fingers enabled them to excel. Other women became engine fitters, motor mechanics and drivers, air controllers, cooks and clerks. As a result, a noteworthy number of male tradesmen — more than 1,000 in August and September 1943 alone — were re-mustered for aircrew training.

* * *

---

* The USAAF commissioned eighty per cent of pilots, 100 per cent of navigators and fifty per cent of bombardiers, but no air gunners, radio operators or flight engineers.

Manipulating human beings is never as easy as it might appear. During the first half of 1943 planners had been faced with a serious shortage of Canadian recruits; in the second half they had to deal with a disconcerting surplus. In August the number enlisting for aircrew training rose to nearly 1,900, and in September there were 2,900. At the same time, the RAF, chivvied by Prime Minister Winston Churchill, was discovering that it had many more aircrew — especially fighter pilots — than it really needed. Moreover, casualty rates were dropping rapidly everywhere except in Bomber Command and the formerly brisk rate of expansion in every sphere of the air war was now stabilizing.

On 16 February 1944, nearly four months before the D-Day invasion of France and more than a year before the end of the European war, a proud but slightly mortified air minister Power told the House of Commons:

> During the three years which followed the signing of the [BCATP] Agreement... the principal preoccupation of Canada and the other partners was to create a training organization on which could be built fighting air forces equal to those of the enemy. Today...this objective has been reached, and we have increasing air superiority...in every theatre of war. ...there are thousands of aircrew being put through Operational Training Units...and further back again there are tens of thousands of young aircrew going through the schools of the BCATP.[18]

He then announced that the Plan would be cut back by forty per cent.

D-Day came and went and the Battle of Normandy hit both the Canadian and British armies hard. Casualties were concentrated in the infantry, where they were much higher than had been predicted despite the Allies' near-total dominance in the air. In Canada, all aircrew enlistment was suspended in mid-June 'until further notice,' while the army struggled to recruit replacements for its lost infantrymen and the Cabinet debated the merits of conscription for overseas service.*

The RCAF had once foreseen a commitment of forty-seven fighter and bomber squadrons for the Far Eastern theatre upon the conclusion of the European war, but Mackenzie King was becoming less and less enthusiastic about such a massive obligation. Moreover, the Americans (who would have had to handle the bulk of their logistics) were not keen to have *any* Commonwealth forces in the Pacific. They could finish that war by themselves. At the *Octagon* conference of the Western Allies, held in Quebec City in September 1944, it was agreed that the RCAF contribution, and the contributions of the other Commonwealth forces, would be diminished by a yet-to-be-determined amount.

Despite the forty per cent cut, there would be more than enough BCATP graduates to man whatever reduced number of squadrons was decided upon and to supply whatever replacements might be required. The Japanese Air Force had never been a really formidable foe. Thus there was no longer any reason to maintain a large training establishment, and in November — just as the Cabinet decided upon overseas conscription for the army — it was announced that the BCATP would be terminated on 31 March 1945. When it ended, the Canadian contribution to the BCATP had been (again in round numbers) some 26,000 pilots, 13,000 navigators and 6,000 air bombers, together with about 26,000 wireless operators and air gunners and 2,000 flight engineers.

As recounted earlier in this chapter, Mackenzie King had initially demanded that all Canadian graduates be posted to RCAF squadrons, most of which were formed overseas. However, many senior RCAF officers, British values and traditions firmly inculcated in them as a result of their First World War experiences (and, in some cases, attendance at the RAF Staff College or service as exchange officers in the inter-war years), had not been unwilling to see RAF aircrew in their units. At the end of 1941, although ninety-four per cent of pilots in RCAF fighter squadrons were Canadian, only forty-three per cent of aircrew in

---

* In June 1940 a National Resources Mobilization Act had authorized conscription for service in Canada; and on 1 August 1942, after a plebescite on the issue, an amending act had empowered the government to send conscripts overseas if and when the Cabinet deemed it necessary.

crewed squadrons could make the same claim. A year later the fliers in fighter units were more than ninety-seven per cent Canadian, but those in crewed squadrons — where a substantial majority of airmen were — still amounted to only 64.6 per cent of the total. By the end of 1943 the fighter percentage had risen by another full point but that for crewed units had dropped by the same amount. Only in 1944 did the figures rise substantially — to 99.5 per cent Canadian among fighter units and 84.2 in all others at the end of the year.

There could have been no complaint if there had not been enough Canadians to make up the necessary numbers; but at the same time many more Canadian boys were being posted to RAF squadrons. Indeed sixty per cent of Canadian aircrew who went overseas would serve in British units. Many of them, especially the commissioned ones, enjoyed the experience (and there was a school of thought that held that 'mixed' squadrons were the best), but a goodly number did not. And, unfortunately, since this is a commemorative history of the RCAF with severe limitations as to its length, their wartime exploits will not be recorded in these pages just as they could not be chronicled in the third volume of the Official History.

Chapter IV

APPRENTICE WARRIORS, 1939–1942

The days of 'bush pilots in uniform' were gone forever. With the coming of the Second World War the RCAF embarked on an exhilarating, if sometimes sorrowful, five-year odyssey that would see it become the world's fourth greatest air force in terms of size and the equal of any in terms of professionalism. The process took time, however, and the first three years were the most difficult. After the war, although it would soon lose that distinction of size, the RCAF would remain the purely military and highly professional force it had become by 1945.

The first wartime patrol was flown on 10 September 1939, the day that Canada declared war on Germany, when a Stranraer flying-boat of 5 Squadron, captained by Flight Lieutenant D.G. Price, set out from Dartmouth, Nova Scotia, on a reconnaissance patrol of the approaches to Halifax. Nothing was seen. Nor was there any sign of the enemy as 5 Squadron provided daytime cover up to the limit of the Stranraer's operational radius — about 400 kilometres — when the first UK-bound convoy left port six days later.

The Stranraers and Wapitis of 5 Squadron were far from being first-line aircraft by current standards, but in November 11 Squadron arrived at Dartmouth with ten brand-new Lockheed Hudsons that were much faster than the Stranraer, were capable of carrying a heavier bomb load (500 kgs), and had a theoretical effective range of nearly 500 kilometres. Of course unlike the Stranraers they could not let down on the water — nor could the Douglas Digbys that arrived with 10 Squadron in April 1940, but these did possess a slightly greater range than the Hudsons.

German submariners had not yet reached the western Atlantic, a voyage that required much time in transit and left too little fuel for operational patrolling. Better U-boats with greater endurance were on the stocks but, for the present, those in service were concentrating their efforts against convoys approaching the British Isles. Nevertheless, rightly fearing the worst,

Eastern Air Command escorted merchantmen and transports leaving Halifax and St. John's and flew occasional patrols through the summer of 1940 and into the fog and gales of the following winter.

When the first opportunity to make their mark did arise, however, 10 Squadron muffed it. In February 1941 the battlecruisers *Scharnhorst* and *Gneisenau* began ravaging shipping southeast of Newfoundland, within range of the Digbys. Two of the latter, en route to provide air cover for one convoy, learned from an armed merchant cruiser below them that a surface attack on another convoy was actually in progress. The air officer commanding, Eastern Air Command, outlined what happened next in an angry informatory letter to the commanding officer at Gander:

> ...both aircraft flew away without bothering to learn the position at which the shelling was taking place and, to make matters worse, proceeded to escort the wrong convoy [already escorted by a capital ship] with the result that several ships of the unescorted *outer* convoy were sunk and the RCAF failed to locate the two large raiders a few miles away.[1]

Having sunk several ships in the scattering convoy, *Scharnhorst* and *Gneisenau* steamed back to their French base unscathed. What happened to the two Digby captains is unrecorded but one doubts if they emerged entirely dismayed.

In the early spring of 1940 the RCAF had three squadrons — one army co-operation and two fighter — serving overseas. The first of them, No 110, was there to work with the army's 1st Division, both division and squadron expecting to join the British Expeditionary Force in France once they reached a satisfactory state of training; however, the fall of France, in June, put a sudden end to that expectation.

For a few memorable months in the fall of 1940 there was the prospect of an invasion of England that would have seen them both go into battle, fully trained or not. However, as winter approached the likelihood of invasion receded, to be replaced by the monotony of training, training and yet more training. For the airmen, even more than the soldiers, this endless loop had a distinct downside — one that was emphasised by their unsatisfactory aircraft.

Army co-operation doctrine at that time, such as it was, envisioned aircraft that could use ordinary grass fields for take-off and landing in order to maintain the closest possible contact with divisional or corps headquarters, but the Westland Lysander — not a small machine — needed larger fields than were common in southern England. And although it was slow enough for effective observation (an attribute enhanced by its high-wing construction) it was neither manoeuvrable enough to evade hostile fighters nor fast enough to outrun them.

Moreover, the squadron was enveloped in clouds of bureaucracy brought about by the need to answer to four different masters. As an RCAF unit it dealt with the newly formed Overseas Headquarters for administrative purposes, while as an army co-operation squadron it was part of the RAF's No 22 Group for training but went directly to Army Co-operation Command for equipment. Adding to the confusion, General A.G.L. McNaughton, as the commander at first of the 1st Division and then of I Canadian Corps, insisted that the squadron came under his operational control.

Each of these competing relationships imposed its own peremptory strains upon the squadron personnel, from the squadron commander down to the lowliest 'erk,' or aircraftman, second class. Morale suffered accordingly and rose only slightly over the summer of 1941 when the Lysanders were replaced with Curtiss P-40 Tomahawks — no great improvement from an operational perspective, since the Tomahawk was a fighter design that was 'obsolete by European standards before the prototype ever flew.' The Tomahawk was half as fast again as the Lysander, but its stalling speed was also much higher, inhibiting precise observation, and it was even less capable of using English meadows for take-offs and landings. It was, perhaps, better for reconnaissance, but it was thoroughly unsuitable for artillery observation, still seen as one of two prime functions of army co-operation flying.

At least the Tomahawk provided dispirited pilots with a chance to entertain themselves with some *very* spirited aerobatics — one such occasion being recounted here in the words of an admiring airman:

Three of our fellows were playing around one day, when one of them takes a notion to fly under a bridge. He came out OK.... The second guy thought that was nothing, so he looped around it, under, then over, then under again. He came out OK. The third guy...shows them both up by doing a slow roll under the bridge — only he didn't quite make it. He lost about 3½ ft [one metre] *of his port wing.... He crippled back to the field though, and found out that his flaps weren't working.... He done the cutest somersault* [on landing] *you ever heard of....*

The bridge in question crossed a highway near the city of Winchester, and the feat is well authenticated.

Another Lysander squadron, No 112, arrived in June 1940, in company with the 2nd Canadian Division, but British doctrine called for one squadron per corps, not one per division. Before the end of the year No 112 had been re-equipped with Hurricanes and converted into a second overseas fighter unit.

No 400 Squadron (together with No 414 Squadron, which by that time had been created to work with II Canadian Corps) was re-equipped again in June 1942 — this time with an early version of the North American P-51 Mustang. The requirement for artillery observation had been abandoned and reconnaissance now seemed to be the only function of army co-operation squadrons. With 'arty co-op' out of the picture and the acquisition of VHF radios, there was no requirement to land anywhere but on a proper airfield. Moreover, 'recce' was made easier by the extensive use of vertical and oblique cameras, so that the pilot's field of downward vision was not as important as it had once been. There was no longer any reason why army co-operation machines should be significantly different from fighters.

These early Mustangs, with their Allison engines, were not as potent fighting machines as the later Packard Merlin-engined versions, but they could out-run any contemporary German fighter at the relatively low levels at which they usually operated, and they could fight, if they had to, with some faint prospect of success, in the hands of a good pilot. Nos 400 and 414 joyfully accepted these new toys and became much happier institutions, although they continued to have difficulties co-ordinating the desires of their various masters.

No 1 (Fighter) Squadron, hastily dispatched to bolster the depleted strength of Fighter Command following the fall of France, had also reached England in June 1940. However, the squadron was far from fit to fight when it arrived. There were new equipment to appreciate and new radio procedures to master, practical medical aspects of high-altitude flying to comprehend and tactical lessons from the French campaign to be absorbed. By 24 August the squadron was assessed as operationally ready and fit to fight in the Battle of Britain, which had already been raging for two weeks and was building towards its climax. Before they were withdrawn for rest and recuperation the Canadians would have made a significant contribution to victory, but Squadron Leader E.A. McNab later described their first operational flight as *'the lowest point in my life.'*

Three four-plane sections were patrolling over Tangmere airfield, providing cover for Spitfires taking off and landing, when McNab spotted three twin-engined machines flying towards Portsmouth, where a raid was in progress. Closing on them, he recognized them for what they were — Bristol Blenheim light bombers — and turned his section away, instructing the other two sections to do the same. His orders were either not heard or misunderstood,* and when the Blenheims fired identifying flares from their Verey pistols the neophyte Canadians mistook them for incoming tracer rounds. A British report of the incident detailed the result in prosaic official language.

At 1640 [4:40 pm] E was approached by Hurricanes and all the Blenheims fired off the recognition signal (Yellow Red). A Hurricane then attacked E and shot the Blenheim down in flames into the sea off Wittering. One body [was] picked up out of the sea by boat, and it

---

* VHF (Very High Frequency) radio was only just coming into service and generally only Spitfires were equipped with it at this time. HF (High Frequency) was not entirely reliable at such close ranges.

is believed that the other member of the crew may have bailed out.

At 1641 AI was attacked by 6 Hurricanes, the first attack damaged the wings, fuselage and starboard engine, and holed the perspex at the front of the aircraft. The Blenheim took avoiding action and fired another [identification] cartridge. A second attack was made by a Hurricane without results. The Blenheim crash-landed at Thorney Island aerodrome with wheels and flaps out of action. The crew escaped with cuts and bruises.

It was not an auspicious beginning but 'the fog of war' is not a meaningless phrase and more than one Canadian would also be the victim of 'friendly' fire before the war ended.

Two days later the squadron was moved to North Weald, where it began the process of redemption. Flying as a wing, together with an RAF Spitfire squadron, No 401 intercepted twenty-five or thirty Dornier bombers escorted by Messerschmitts. While the Spitfires took on the fighters the neophyte Canadians shot down three Dorniers and damaged three more, while losing three machines (and one pilot) of their own.

They continued to take a toll of the enemy — but they also took quite heavy casualties of their own. By the end of the first week in October the squadron had lost three killed and eight wounded out of the twenty-one original pilots and the six reinforcements it had received from 110 and 112 squadrons. Their RCAMC doctor — the RCAF still had no doctors of its own, never mind flight surgeons — noted what battle fatigue was doing to the pilots:

There is a definite air of constant tension and they are unable to relax.... This constant strain and overwork is showing its effects on most of the pilots, and in some it is marked. They tire very easily and recovery is slower. Acute reactions in the air are thereby affected.... Needless casualties are bound to occur as a result of these conditions, if continued.

Somebody was paying attention. As the battle died down — its formal conclusion came on 30 October — the squadron was taken out of the front line and posted to Scotland. In December it became part of the defences of the Royal Navy base at Scapa Flow, in the Orkney Islands, where, their tranquillity disturbed only by high-flying German photo-reconnaissance machines operating well above their Hurricanes' ceiling, the squadron diarist found occasion to complain:

...all accommodation is very cold and not good. Thurso is a very small village with little entertainment, few women and the coldest hotel rooms ever experienced. Dispersal is in an old dilapidated farmhouse. It is dark until 0915 hours and the sun goes down about 1530 hours and never gets very high up in the sky.

On the other hand, there was a cornucopia of good Scotch whisky which the diarist failed to mention!

By mid-February 1941 it was time to move again — south to Digby, in eastern England, and re-equipment with Hurricane IIs. The day-fighter war was about to enter a new phase, in which the RAF and RCAF would turn to the offensive.

British bombers had been on the offensive, after a fashion, since the middle of May 1940, when, in a vain attempt to constrain the *Blitzkrieg,* Bomber Command had begun attacking oil refineries, factories and railway yards, with very limited success. 'We drove through many of the Ruhr centers...the Allies were supposed to have bombed...the last few nights,' a then-neutral American reported on 19 May.

We naturally couldn't see all the factories and bridges and railroad junctions...but we saw several, and nothing had happened to them. The great networks of railroad tracks and bridges around Essen and Duisburg...were intact. The Rhine bridges at Cologne were up. The factories throughout the Ruhr were smoking away as usual.

Indeed it would be two more years before the strategic bombing offensive began to make a significant mark on German cities, and three before it attained the stage of consistent mass destruction. But, as we shall see, from that time on until the end of the war Allied bombing would have a terrible impact on both industry

*No 1 Squadron works up on Hurricanes, Vancouver, 1939.
(CFP, RE 23088)*

*Squadron Leader E.A. McNab in Hurricane, No 1 Squadron, September
1940 during the Battle of Britain. (CFP, PMR 71-536)*

and people. Again in round figures, 13,000 tonnes of bombs were dropped on Germany in 1940, 32,000 tonnes in 1941, 48,000 in 1942 (when the USAAF joined the offensive), *four times as much* in 1943, and *nearly a million tonnes* in 1944.

The first of fifteen Canadian heavy bomber squadrons was No 405, formed overseas (like all its fourteen peers) in April 1941. Initially, the squadron flew twin-engined Vickers Wellington Mk IIs, exceptionally sturdy machines (which was just as well, given their dangerous utilization) with a crew of six, a maximum speed approaching 400 kph, an operational ceiling of about 4,500 metres, and the ability to drop as much as 2,000 kilograms of bombs on targets up to 1,400 kilometres away from their base.

The Wellingtons were British designed and built; so was No 405's squadron commander, since the RCAF had no pool of experienced bomber crews from which to draw. The two flight commanders also came from the RAF, although they were both Canadians. When the squadron formed, only 16.5 per cent of the aircrew were Canadian, and more than half of that contingent were Canadians serving in the RAF. By August, however, forty-five per cent would be from the RCAF, and by late fall the RCAF content would rise to fifty-three per cent.

The squadron flew its first four sorties on the night of 12/13 June 1941, against the railway marshalling yards at Schwerte. One machine turned back before reaching the target, complaining of engine problems. The others claimed to have bombed the target, '*bursts being seen and fires observed,*' but admitted that '*results were difficult to assess owing to ground haze.*'

The next operation came three nights later, when six aircraft (together with ninety-eight from RAF units) were sent to bomb the main railway station at Cologne. This time, one RCAF machine failed to return and a second was badly damaged by a German night-fighter, while the raid was certainly not a success. Germans authorities recorded only fifty-five bombs falling on the city — the bomb loads of a dozen machines — and damage was negligible. Results were perhaps even poorer on 22/23 June, when three of seven 405 Squadron aircraft participated in an attack directed against Wilhelmshaven. The only victims were residents of a small village some distance from the port, where one house was hit.

Pre-war doctrine favoured day bombing but exorbitant casualty rates had dictated a conversion to night bombing, particularly since night-fighters, operating without the advantage of airborne radar at this stage of the war, were not the major threat they would later become. However, accurate navigation on long flights over smog-ridden and unlit regions was proving much more difficult than anyone had expected.

* * *

In May 1941 Eastern Air Command received the first firm intelligence (acquired through radio triangulation) of a U-boat within air range of Newfoundland. Two months later an RCAF Consolidated Catalina — the first of four — flew into Botwood, the flying-boat base in southern Newfoundland. The Catalina (the RCAF version became known as the Canso) was a graceful, twin-engined, high-wing flying-boat with a cruising speed of 185 kph — more than enough to escort a convoy or attack a U-boat, and there was no question in the western Atlantic of having to run from German fighters!

More importantly, the Catalina had an airborne endurance of *twenty-five* hours and a theoretical operational radius of nearly 2,000 kilometres while carrying four depth charges in racks under the wings. However, the real world was quite different from the theoretical. In practice, twenty per cent of the Catalina's fuel needed to be reserved in case it became necessary to reach the nearest alternate base at North Sydney, Nova Scotia, while a reasonable minimum of time on station was eight hours. Even flying back to base against the prevailing winds the Catalina might consume twenty per cent more fuel if the wind picked up, so that the *effective* range was no more than 800 kilometres.

In October 1941 a 'wolfpack' of U-boats began to gather just to the east of the Strait of Belle Isle, between the island of Newfoundland and southern Labrador, and on the 25th of the month Eastern Air Command made its first sighting of, and attack on, an enemy submarine. The pilot of the Digby involved, Squadron Leader C.L. Annis, takes up the story.

At approximately 1450 hours...I sighted a submarine.... Only its conning tower was visible and it disappeared into a wave as I watched. The vortex of its dive was plainly visible and the shadowy darkness of its hull showed for a few seconds. As the vortex and bubbles built up towards the east, I was able to decode what had been troubling me all along — the direction it was moving and therefore at which point to aim in the attack.

By this time, which I should judge to be 20–30 seconds after first sighting, we were in a 30–40 degree dive as I turned to the left...to make a quartering astern attack. Remembering to aim short and ahead, and estimating a six-second interval between release and detonation, I released the bombs in salvo, by means of the pilot's emergency release, when at a little less than 300 feet [90 metres] *indicated on the altimeter, and in an angle of dive of approximately 20 degrees.... The strong wind...had caused me to undershoot somewhat.*

Undershooting made no difference, since someone had switched the bomb-release lever back to the safe position at some point during the flight. That U-boat, like *Scharnhorst* and *Gneisenau* before it, sailed back to its French base unscathed.

After that first contact the weather worsened, flying becoming impossible on many days. During one such stormy interval at the end of the month, eleven U-boats banded together to attack an eastbound convoy well within range of RCAF cover. Frustrated airmen prayed for the weather to improve, but before it did four ships had been sunk and the convoy ordered to return to Canada through the Strait of Belle Isle — where two more ships were grounded in fog. Towards the end of November 1941 further deterioration in the weather compelled the Catalina detachment at Botwood to withdraw to North Sydney.

On the other side of the Atlantic the RCAF was re-numbering its overseas squadrons to avoid potentially embarrassing (or even dangerous) confusion. With identically numbered squadrons it would be easy to confuse RCAF units with those of the RAF, RAAF or RNZAF; so, by common consent, the Dominions were allocated blocks of numbers above 400 — RCAF squadrons were to be numbered between 400 and 449, RAAF and RNZAF between 450 and 499.

Of the Canadian squadrons already overseas, No 110 had been re-numbered as 400; No 1 became 401, No 112, 402. The first squadron to be *formed* with a 400-block number, on 1 March 1941, was another fighter unit, No 403; then came the first Canadian squadron in Coastal Command, a 'maritime fighter' unit formed on 15 April.

No 404 Squadron was initially equipped with a variant of the twin-engined Bristol Blenheim IV, designed (and more often used) as a light bomber, but in this case armed with a pack of four forward-firing .303 machine-guns under the fuselage, an addition that apparently justified calling it a long-range fighter. The squadron's usual assignment was to guard coastal convoys off the east coast of England against marauders, although their Blenheims had 'neither the armament nor the speed to give combat on anything like even terms to the Focke-Wulf [200] or the Heinkel He 111 [bombers]....' They would have been at even more of a disadvantage against enemy fighters. Luckily, Luftwaffe incursions were becoming relatively rare; in the aftermath of the Battle of Britain few Germans were anxious to encroach on British airspace or adjacent coastal waters by daylight.

The next Canadian contribution to Coastal Command (which was always the RAF's poor relation in terms of equipment, running a distant third to Fighter and Bomber commands) was No 407, formed in May 1941 as an anti-shipping unit — the opposite side of No 404's coin, or attempting to do unto the enemy that which he was largely unable to do unto us! To that end, the squadron was issued Lockheed Hudsons, which, as we have noted, had an acceptable turn of speed and the ability to deliver 500 kilograms of bombs. But what was really needed was flak-proof armour plating.

Daylight strikes, which were the norm in 1941, were often extremely costly, as an RAF squadron assigned to similar duties had already discovered; but 407 Squadron was lucky enough to start off with a British commanding officer who, in effect, turned it into a dusk- or night-attack unit, depending on the weather and the current phase of the moon. Heavy overcast or no moon meant dawn or dusk attacks in low light; clear, or partially clear, moonlit skies enabled the squadron to strike at night.

In the first six weeks of operations eleven strikes on enemy shipping off the Dutch coast were made, without any losses — although successes were not plentiful, either. Damage might be caused by a near miss, but sinking a ship usually required a direct hit and ships were not easy targets even from low level. Casualties then began to mount as flak defences improved and enemy gunners increased their vigilance between sunset and dawn. By the end of January 1942 the squadron had made eighty-two attacks and lost twenty-nine aircrew, as well as the thirteen gallant groundcrew who were killed when a crash landing sent them rushing to help any survivors just as the bombs on board exploded.*

We were always waiting to go out and it didn't pay to have too much imagination while you were...just waiting for orders to attack a well defended convoy....

You would go over the moves you were going to make. Where was the moon? If the convoy was on such and such a course as reported, what would be the best run-in for an attack, and what about getting away after it was over? If it had a destroyer escort as well as flakships, how would it be formed up?...

Patrols along the shipping lanes were not as frightening as the strike, for you were busy until the convoy was located, and then the action happened so fast that it was over almost before you knew it.... If you were careful to attack from the dark side and stay well down, you might avoid having to make your run-in through a curtain of flak.[2]

---

* Regrettably, all air-war histories (including this one) tend to downplay or ignore the part played by groundcrew and the danger they sometimes faced in peace as well as war. Their story was rarely as dramatic or as exciting as that recounted here, but their work was every bit as essential. Without them of course there would have been no air war.

The worst was yet to come, as the flak got heavier month by month. Attacks had to be made from low level to have any hope of success and were always within range of multiple, automatic flak. By the end of the war, anti-shipping strikes would have proved to be the most dangerous thing in which a Second World War airman could participate.

With the Battle of Britain won, the fighter boys began, in the words of their commander-in-chief, 'leaning forward into France' — and by mid-summer 1942 there were twelve RCAF squadrons in Fighter Command. Three were night-fighter squadrons, whose life was relatively unexciting, and one — No 418 — was an Intruder unit.** The other eight day-fighter units were among the sixty-odd squadrons that carried out the Command's bizarre and costly offensive, apparently devised to give fighter pilots something to occupy themselves with now that they had secured the airspace over Britain.

'Leaning forward' involved three kinds of leaning. 'Rhubarbs' were unsupervised lunges across the Channel by two or three machines, usually flown at the pilots' own discretion and simply seeking targets of opportunity, either in the air or on the ground; organized forays by larger formations — squadrons or wings — were called 'Rodeos' and were deliberate attempts to lure the Luftwaffe into battle; while even larger formations — sometimes involving half a dozen wings — that escorted light bombers (simply to increase the level of irritation) were labelled 'Circuses.'

While the raiders always enjoyed a numerical advantage, the tactical edge lay with an enemy who not only could choose exactly when and where to fight but, for much of the time, retained a technological advantage as well. The Me 109F could fly higher than the Spitfire VB and had better overall performance. 'At all heights a Spitfire can turn inside a...109 but the 109 appears to have quicker initial acceleration in a dive and also in climbing,' reported RAF

** Intruders were long-range night-fighters whose business it was to patrol the vicinity of German airfields in the hope of catching unawares German night-fighters taking off or landing. No 418 was Canada's only Intruder squadron.

Hercules on a LAPES drop. (CF photo, ENC 82-467)

◀ *Wellington bomber in England. (CF photo, PC 2475)*

A Vampire, the RCAF's first jet fighter. (CF photo, PC 251)

CF-100 banking. (CF photo, PCN 5599)

T-33 Silver Star at Weapons Practice Unit, Cold Lake. (CF photo, PCN 209)

Dakota supporting Exercise MUSKOX, April 10, 1946. (CF Photo, ZK 1093-8)

Comet jet transport in flight. (CF photo, RNC 164)

Boeing 707 refuels a CF-5 in flight. (CF photo, REC 85-4520)

Labrador helicopter over Rocky Mountains. (Larry Milberry and Canav Books)

Labrador makes a rescue. (CF photo, SEC 78-1533)

Relief work during 1998 ice storm. (CF photo, IAC 98-004-7A)

Kiowa Helicopter on patrol over Montreal, 1976 Olympics. (CF photo, IMOC 76-844)

Caribou on United Nations work. (CF photo, PC 2696)

Search and Rescue Canso. (CF photo, RNC 45)

Piasaki and Albatross. (CF photo, PCN 4346)

Aurora maritime patrol aircraft. (CF photo, REC 85-150)

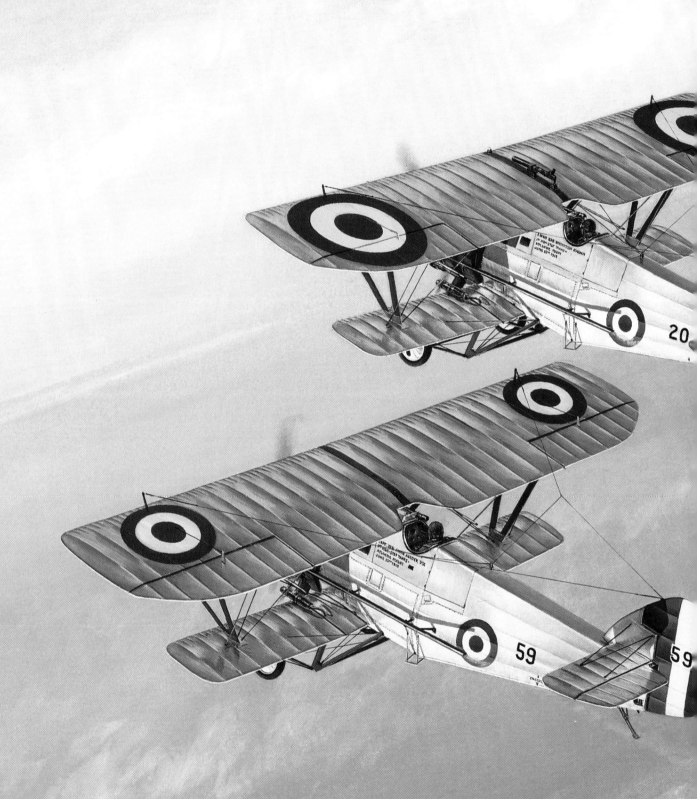

RCAF Golden Hawks over Niagara Falls. (CF photo, PCN 1045)

Golden Centennaires. (CF photo, PCN 67-24)

The Sky Lancers aerobatic team, 1954. (CF photo, PC 1038)

The Snowbirds. (CF photo, PCN 72-65)

CF-101B Voodoo fires a practice rocket at night. (CF photo, BNC 72-2844)

CF-101B intercepts a Soviet BEAR. (CF photo, PMRC 84-940)

Pinetree Line radar site, Lac St. Denis. (CF photo, RNC 90)

Sabres of 421 Squadron, June 1957. (CF photo, PCN 109)

CF-104 flies past by a German castle. (CF photo, PCN 75-364)

Four CF-18s on Operation RHINE PROSIT, January 1993, the last NATO exercise by Canadian fighters before their return to Canada from Europe. (Larry Milberry and Canav Books)

Lancaster bomber in England. (CF photo, PC 2513)

Mitchell bomber of 418 (Auxiliary) Squadron during the 1950s. (CF photo, RNC 27)

Harvard, student and instructor, August 1956, Penhold, Alberta. (CF photo, PC 1481 (N))

Canadian Warplane Heritage Lancaster and CF-18.
(CF photo, AEC 89-778)

Spitfire XIV, 421 Squadron, Petit Brogel, Belgium, January 1945.
(CF photo, REC 89-1492)

*Commonwealth Air Forces Memorial, Green Island, Ottawa.
(CF photo, PCN 1346)*

Intelligence. But although tight turns were an advantage in dog-fights, enemy pilots usually chose to avoid such clashes and dive straight through their opponents' loose formations, firing bursts of 20-mm cannon fire at one or two machines as they flashed by.

When the apparent objective of the invaders became attacks on such massively built, 'sensitive' targets as power stations, which demanded weightier ordnance, Circuses became 'Ramrods,' involving one or more heavy bombers. Pilot Officer Hugh Godefroy of 401 Squadron, flying his first Ramrod in July 1941 (just eleven days after Hitler attacked the Soviet Union and gained the Commonwealth its first major ally), *'was startled at the number of aircraft in the formation. Above us were the Spitfires, a squadron at every thousand feet up to twenty-seven thousand.... Below us, like a mother hen with its brood, was a single four engined [Short] Stirling bomber surrounded by squadrons of Hurricanes.'*[3]

October brought the first success of a notable career when Pilot Officer R.W. McNair shot down his first enemy aircraft towards the end of a Circus over the Pas de Calais.

[Finding]...a group of seven E[nemy] A[ircraft] circling a pilot in the sea, I picked out one, opened fire at him at about 250 yds [225 metres], *a quarter astern; he went into a sharp left-hand diving turn. I got on his tail and gave him a 3-second burst, closing to 60 yds* [54 metres]. *I overshot him, pulled away to the right, and in going down I saw him go straight into the sea.*

Minutes later McNair was shot down himself, although he also managed to damage his opponent. *'Flames were now coming out of my cockpit, so I put my nose down. Finding that my engine was cutting out, I pulled up to 400 feet* [120 metres] *and baled out into the sea. I was picked up about 15 minutes later.'*

McNair would be credited with sixteen victories — most of them achieved over Malta, where he served in an RAF squadron during 1942 — and reach the rank of wing commander before the war ended. Only Squadron Leader H.W. McLeod, killed in action over Germany in September 1944, would shoot down more enemy machines while serving in the RCAF,* most of his nineteen victories also coming while he was serving in an RAF squadron on Malta.

'Leaning forward' was an expensive way to make war. In the six and a half months between 14 June and 31 December 1941, Fighter Command lost 395 pilots killed, taken prisoner or missing on such missions. A total of 731 German aircraft were claimed to have been destroyed, but actual losses attributable to RAF/RCAF action were only 103 — and almost certainly the pilots of more than half of those shot down lived to fight again. The five RCAF squadrons involved claimed twenty-two destroyed, with another fifteen 'probables,' while losing twenty-one of their own pilots; but it is impossible to separate German losses into those brought about by RAF action and those that should be credited to the RCAF.

A winter lull in the fighting was followed by a renewed and reinforced (by ten more squadrons) 'lean' in the spring of 1942, timed to coincide with the first major Soviet offensive of the war, which was expected to draw numbers of Luftwaffe fighters and fighter pilots to the Eastern Front. Some squadrons did indeed move east; but those that were left behind, fighting on their own terms and increasingly equipped with the newest German fighter, the Focke-Wulf 190, still more than held their own.

Between the beginning of March and the end of June, Fighter Command admitted the loss of 259 aircraft and pilots, while claiming 197 German aircraft destroyed. Actual German losses were only fifty-eight, with perhaps twenty-five pilots killed or incapacitated. The five RCAF squadrons (all now flying Spitfires, since No 401 had converted from Hurricanes in the fall of 1941 and No 402 in the early spring of 1942) lost twenty pilots between them, while claiming only nine victories. Again, it is impossible to allocate German losses between the RAF and RCAF.

---

* The top-scoring Canadian fighter pilot of the war was George 'Screwball' (or 'Buzz') Beurling, but only the last two of his thirty-one credited victories were gained while he was with the RCAF. As with McLeod, most of his successes were achieved while serving on Malta. It must be remembered that the calibre of opposition there was not comparable to that of north west Europe, where the Germans concentrated their best airmen.

However, for all this vast imbalance in losses, the Anglo-Canadian fighter forces were better able to cope than the Luftwaffe was. British factories were churning out more than enough replacement aircraft and the BCATP was producing all the pilots required, while German reserves of men and machines were being eaten up on the Eastern Front, in North Africa and in training accidents. New German pilots were not as well trained as they might have been and the Luftwaffe was coming to rely on a small cadre of extraordinarily skilful veterans.

Turning to the Canadian night-fighter squadrons, all three were formed in the spring and summer of 1941. Nos 406 and 409 had the good fortune to be equipped with Bristol Beaufighters almost immediately, but a hapless 410 Squadron languished on obsolescent Boulton-Paul Defiants until the following summer and would not be completely re-equipped with Beaufighters until the summer of 1943.

The two-seat, twin-engined Beaufighter was a potent piece of work, capable of 600 kph in a 'clean' configuration and armed with four 20-mm cannon in the nose and six .303 machine-guns in the wings. The antennae required by the AI (airborne interception) radar had an adverse effect on the Beaufighter's speed but still left it fast enough to overtake almost any contemporary bomber. The enemy posed little threat in return, but operating high-performance aircraft in darkness, often during poor weather and before the development of blind-landing systems, was a dangerous undertaking in itself. During their first year of operations the three squadrons suffered fourteen fatalities in various accidents.

Since the AI radar had a very limited range, contacts were not common. No 406, at Acklington in northern Yorkshire, enjoyed the best location, being perfectly situated to intercept raids on the industrial northeast of England (a favourite Luftwaffe target since relatively little time had to be spent in British airspace), and, not surprisingly, experienced the most success. The other two squadrons were based well away from the action and had little or no luck. By June 1942, No 406 had registered five victories and No 409 one, but the unfortunate No 410, flying those awful old Defiants, had not scored at all.

The climax of the fighter offensive came with the ill-fated Dieppe Raid of 19 August 1942. Whatever the motivations and intentions of the War Office, Lord Louis Mountbatten's Combined Operations Command and General McNaughton's First Canadian Army, the Air Staff saw it simply as a greater-than-ever opportunity to compel the Luftwaffe to give battle. A Ramrod par excellence.

The RAF/RCAF order of battle mustered fifty-six fighter squadrons, including forty-eight Spitfire squadrons — four of the RAF units flying the new Spitfire IXs, which were a match for any German fighter. All five Canadian fighter squadrons participated, each flying two or three missions through a day that brought probably the greatest dog-fight of the war. In that kind of confused mêlée the Spitfire Vs of 412 Squadron could hold their own against the Focke-Wulf 190.

There were FW 190s all over the place around 2,000 feet, and we were the only Spits at our height. Some 190s started to dive down on the Hurr[[can]es. We tore after them and they, seeing us coming, started to break away. Just then, someone yelled 'Red Section, break.' There were some 190s on our tail. We went into a steep turn to the right and shook them off. I lost the others for a few seconds. The Flak started to come up at us in great volume.... I spotted my No 1 and joined him. Just then the C[ommanding] O[fficer] yelled 'Let's get out of here'.... We landed with the whole squadron intact.

Nos 400 and 414 — now called 'fighter reconnaissance' squadrons — both got into the action, keeping a wary eye open for German reinforcements moving towards the battle. Flying Officer H.H. Hills, in his Mustang, spotted two Focke-Wulf 190s when 'we were at nought feet.'

The enemy aircraft turned and followed us, holding his height.... I turned in...and gave the leader a short burst.... By this time the Focke-Wulf behind me had opened fire on me, but his fire was passing on the port side.... I slipped violently to starboard, towards the ground.... The Focke-Wulf passed me and started a steep turn.... I followed and gave him a 2-second burst.... I observed parts flying off and an explosion about one foot [30 cms] *behind the engine cowling.... I gave him another 2-second burst....*

Fuelling and bombing up a Wellington of No 405 Squadron, 1941. (CFP, PL 4501)

Hudsons of No 11 (Bomber Reconnaissance) Squadron, 3 September 1941. (CFP via N.S. MacGregor, PMR 77-393)

Beaufort, Patricia Bay. (NAC, PB 1405)

...I looked back at the Focke-Wulf I had hit and it went into a grove of trees with dense black smoke coming out of the aircraft. There was no explosion when he hit.

Hill's leader was shot down by the other enemy machine but ditched his Mustang in the sea alongside a British destroyer and was rescued 'without getting his feet wet.'

Losses on each side were proportioned about as usual, eighty-nine RAF/RCAF to twenty-three Luftwaffe, or roughly four to one, but by nightfall only seventy of the 230 German fighters that had been serviceable that morning were still fit to fight. Hasty repairs and the issue of all reserve machines brought the number up to 194 on the 20th, but for the time being 'there were no further reserves available.' On the other side of the Channel all losses were made up by the next morning and there were still many reserve aircraft and replacement pilots in hand.

Heavy bombers took many more man-hours to build than fighters, and bomber crews were complex, multi-person organizations, requiring careful co-ordination of resources to produce. Consequently, the bomber arm expanded rather more slowly than the fighter force. Any difference in growth rate, however, was more than made up by the introduction of heavier bombers carrying greater bomb loads.

The second RCAF bomber squadron to form was No 408, followed by the end of 1941 by Nos 419 and 420. All three were initially equipped with twin-engined machines, as was the usual Bomber Command practice. 'Promotion' to four-engined Handley-Page Halifaxes or Vickers Lancasters came with seniority. Seven more squadrons formed in 1942, but only No 425, the specifically French-Canadian squadron, was formed in time to join Operation MILLENNIUM — the first thousand-bomber raid, directed against Cologne on the night of 30/31 May 1942. (There would be other thousand-bomber raids, but they never became commonplace and none was as successful as this first one.)

The year 1942 brought the introduction of *Gee*, the first electronic navigation aid, which, when it was working properly (and that was not all the time), could theoretically place a bomber within a roughly diamond-shaped area five to six kilometres long by one kilometre wide. *Gee* depended on line-of-sight radio waves emanating from southern and eastern England, and the curvature of the earth combined with the operational ceiling of the bombers to limit its range. Nevertheless it had some impact on target-finding when it was first introduced, the percentage of crews dropping their bombs within three miles of the aiming point rising from less than twenty-five to over thirty between April and June.

Unfortunately, by then the Germans had discovered how to jam *Gee* transmissions and bombing accuracy quickly dropped back almost to its former level. Not quite to its former level, however, for the jamming of *Gee* coincided with the introduction of Pathfinders — specially selected crews whose job it was to lead the way and mark targets (with differently coloured flares) for the main bomber force following on behind. That was a refinement that the Germans could not counter. But precision bombing by night was still out of the question.

Long before you reached the target area you would see ahead of you a confusing maze of searchlights quartering the sky, some in small groups, others stacked in cones of twenty or more. These often had a victim transfixed, as if pinned to the sky, their apex filled with red bursts of heavy flak. The ground would soon be lit with lines of reconnaissance flares like suspended street lights.... As the raid developed, sticks of incendiaries criss-crossed the ground, sparkling incandescent white until a red glow would show the start of a fire.

The Germans liberally sprayed the ground with dummy incendiaries and imitation fire blocks in the neighbourhood of important targets, hoping to attract a share of the bombs. Gun flashes, photoflashes, bomb-bursts, streams of tracer of all colours, and everywhere searchlights — it was all very confusing....[4]

Losses to flak and night-fighters in Operation MILLENNIUM were 3.9 per cent, but overall loss rates were running at about five per cent during the summer of 1942, and that was approaching the limit of sustainability. 'A strategic bomber force would become relatively ineffective if it suffered operational losses of 7 per cent over a period of three months,' the Air Staff calculated, 'and the operational effectiveness may become unacceptably low if losses of 5 per cent were maintained over that period.' Even a five per cent rate meant that only one crew in five could expect to survive a normal tour of thirty missions.

Such conclusions (and the statistics behind them) were strictly classified information, and loss rates were not encouraged as a topic for conversation on bomber stations, where a refusal to fly any more sorties was severely punished (at least in the case of non-commissioned aircrew). Nevertheless everyone could see for himself, by the sudden disappearance of others from the mess, that his chance of completing a tour was poor, and he knew that only if he was lucky would he become a prisoner of war. In late April, Flight Sergeant Douglas Alcorn, a navigator in 418 (Intruder) Squadron, visited a friend serving in a Canadian bomber squadron and accompanied him to the local pub.

I found the atmosphere in the place far from hilarious or even joyful. As a matter of fact, a lot of the lads were sitting there just staring into space until they'd drained their glasses, at which time they raised their fingers to signal another pint.... Perhaps it was the difference between the bomber and the fighter role, the long nightly lists of casualties compared to our lighter losses, the flights right into the German heartland compared to our nuisance flights around France and the Low Countries. It was probably a lot of other things as well, too complicated to explain, but still visibly there and gnawing away at these young boys at their tables around us. Ev finally suggested we shove off and I heartily agreed.[5]

The problem of Canadianizing the squadrons overseas was not going well either. At the end of 1941 there were 6,700 Canadian aircrew serving overseas, nearly all in the United Kingdom, but only 600 of them were serving in RCAF squadrons, where they formed fractionally more than fifty per cent of the aircrew. In December 1941 Air Vice-Marshal L.F. Stevenson, the air officer-in-chief overseas, thought that efforts to alter that balance were inadvisable:

> The best squadrons are mixed squadrons. Every man has something to give, if you put them together they pull. Much better results are achieved by mixing the men.... Canadian aircrews in England are operating under very highly skilled staffs. Any weaklings are tossed out. There is absolutely no mercy about it. The Canadians are well looked after by RAF men with two years' war experience.... They are working hard, fighting hard, and doing a great job.

Some of the aircrew concerned agreed with him, but many did not. Certainly a majority of other ranks heartily disliked the class system rampant in the RAF, not to mention the broader question of being labelled 'colonials'. It was, given the approach taken by the Canadian government, inevitable that a majority of RCAF aircrew would have to serve in RAF squadrons, but it was surely not necessary that RCAF squadrons should have such a large proportion of RAF personnel. Nor did that quota of RCAF officers who now had sufficient experience to justify their promotion to flight or squadron commands appreciate a situation that denied them their chance.

Back in Ottawa, air minister C.G. Power had begun to glimpse the political dangers of neglecting Canadianization. In October 1941 Stevenson was replaced by Air Marshal H. Edwards, who took a much more nationalistic view. With Edwards's arrival in London the situation began to improve, but only slowly. It was hard to pin down who, exactly, was responsible, reported his deputy, Air Commodore Wilf Curtis:

> In spite of the very evident desire to cooperate toward achieving this end on the part of both Air Ministry and this Headquarters, there have been many recent examples of postings which have the effect of postponing, rather than advancing, the date of arriving at a condition under which all aircrew positions in RCAF squadrons would be filled by RCAF personnel.
>
> There are RCAF officers with considerable operational experience who are considered competent to fill the squadron and flight command vacancies in the newly forming RCAF squadrons, yet apparently due to the fact that recommendations for postings are frequently made at the Group level, RAF personnel are posted to positions in RCAF squadrons at considerable incon-

venience to the RAF, when, in actual fact, eligible RCAF personnel are available in other Groups.

Nevertheless with Edwards driving the process the situation began to improve. By the end of June 1942, seventy per cent of aircrew in RCAF squadrons were Canadian — ninety-six per cent in fighter squadrons. At that time, too, the fifth RCAF bomber squadron, No 425, was formed, the first to be formed under an RCAF commanding officer — although by that time Wing Commander J.E. Fauquier, who would subsequently become Canada's outstanding bomber leader and, indeed, one of the greatest in the Allied air forces, had been commanding 405 Squadron for four months.

Chapter V

JOURNEYMEN CAMPAIGNERS, 1942–1943

During 1942 German submarines moved further west in larger numbers, even into the Gulf of St. Lawrence, where *U-553* — 'the first enemy warship in those waters since Canada had become a nation seventy-five years before'[1] — sank two small steamers off the Gaspé coast on 12 May.

Finding the enemy from the air was more difficult than one might think. All too often the Royal Canadian Navy failed to pass timely intelligence based on decrypted U-boat signals, while Eastern Air Command could rarely muster more than thirty serviceable aircraft (out of an establishment of fifty) to patrol the 140,000 square kilometres of the Gulf. Moreover, EAC's obsolescent thirty-centimetre radars could discern a conning tower at something more than eyeball range only in calm seas and not at all in heavier weather. And daylight or dark, eyeball or radar, it was quite possible to confuse a partially surfaced U-boat with a breaching whale, whales being much more common in the Gulf then than now.

Between May and October EAC flew nearly 1,600 sorties over the Gulf and made seven attacks on U-boats (and two on whales) but sank none. On those rare occasions when a U-boat was attacked, a want of sufficiently powerful, shallow-set depth charges was responsible, at least in part, for their inability to sink it. Their approaches may also have left something to be desired, but EAC's Amatol depth charges were about thirty per cent less powerful than the new Torpex charges that Coastal Command had just begun to receive on the other side of the ocean. Not until November would the Canadians get Torpex charges, and a detonator that could set them off at a depth of only five metres rather than the earlier figure of seventeen metres would not be available until early 1943.

Twenty-one ships (nineteen merchantmen and two naval escorts) had been sunk by September 1942 when the Gulf was temporarily closed to ocean-going shipping — two months before ice would have closed it in any case. Coastal vessels continued to ply the Gulf, however, and in October the Sydney–Port aux Basques ferry, *Caribou,* sailing without air cover, was sunk with the loss of 136 lives.

However, air cover was gradually becoming more widespread and intense as both the groundcrews and aircrews of EAC were getting better at their jobs. Shortly after sinking *Caribou,* the commander of *U-69* signalled his headquarters that 'constant patrol by aircraft with radar' was complicating his mission and reducing the likelihood of further successes. Other U-boats were passing similar messages and Admiral Dönitz noted at the end of the month that *U-43* patrolled 'for seventeen days and operated against two convoys without success.' That was not as spectacular a success for the Gulf airmen as the sinking of *U-43* would have been, but it was every bit as valid an accomplishment. Suppression, in anti-submarine warfare, was almost as good as a sinking, even if it was less satisfying.

Crews patrolling off Newfoundland and Nova Scotia learned to fly in conditions well below the minimum tolerated in the United Kingdom's Coastal Command. The mixing of warm Gulf Stream water with the icy Labrador Current produced almost perpetual fog over the Grand Banks, and for at least half the year there was the threat of icing to contend with. To the south of the Banks, however, conditions were marginally better and it was there that EAC recorded its most sensational achievements.

After a number of sightings and several unsuccessful attacks on U-boats off the Nova Scotia coast, on 31 July 1942 a Hudson of 113 Squadron piloted by the squadron commander, Squadron Leader N.E. Small, was:

> ...on a special sweep at an altitude of 3,000 feet [900 metres] in response to fresh intelligence. The weather was ideal, a slight summer haze making visibility poor from the surface of the water. Three miles ahead the U-boat quietly cruising along, was taken quite unawares as the Hudson dived to the attack. Sailors were seen scrambling for the hatch, and most of the boat was still visible when the depth charges went tumbling down around it.... On the third circuit the front gunner opened fire when the conning tower briefly reappeared. Large air bubbles continued to surface until a heavy underwater explosion brought a large quantity of oil swirling up to mark the grave of *U-754* — Eastern Air Command's first kill.

Just hours later another No 113 Hudson attacked *U-132,* and Small also attacked *U-458* on 2 August and *U-89* on 5 August. Even when unsuccessful, such strikes made the U-boat commanders go about their business in more cautious fashion and thereby reduced the threat to shipping.

Nevertheless, in October a visitor from Coastal Command found much to criticize. 'Generally speaking,' he reported, 'Eastern Air Command is a very long way behind any other place I visited in either Canada or the United States.' Although solving the Enigma cyphers that the U-boats used had enabled the Admiralty (and Naval Service Headquarters in Ottawa) to forecast with considerable accuracy just which convoys were going to be attacked, and where, the Canadians were often overextended. Too great a proportion of resources was being devoted to close escort of every convoy. It would be better, he advised, to concentrate attention on those convoys that signals intelligence suggested were to be attacked and range further out from them, especially in the early morning and late evening hours. Finally, EAC aircrews should patrol at greater height — 600 to 800 metres, giving them a broader picture, would be more effective than the 300 to 400 they currently favoured.

The new tactics soon proved their value. On 30 October a Hudson of 145 Squadron, responding to 'fresh intelligence,' sighted a conning tower breaking the surface:

> ...at the time the depth charges were released the U-boat was almost fully surfaced. Four 250-lbs [113 kgs] Mk VIII depth charges with Mk XIII pistol [detonators] set to 25 feet [7.5 metres] at an angle of 30 degrees across the U-boat from port astern to starboard bow. All the charges functioned correctly and explosions were noted bracketing the U-boat, the centre two charges on opposite sides of the hull and very close to it.

The explosion raised the U-boat in the water and 60 feet [18 metres] of its stern raised on an angle of 40° to the horizontal. The U-boat then settled and a large oil slick and air bubbles merging with the rough sea appeared immediately.

Another U-boat fell victim to a Digby of 10 Squadron later that evening, and an attack on a third, made the next day, was assessed as 'an excellent attack deserving of more concrete evidence of damage.' The only evidence recorded was streaks of oil, 'probably squeezed out of the compression valves on the U-boat's fuel tanks.'

Canadian squadrons accounted for half the U-boat kills in the northwest Atlantic during 1942, with two successes going to aircraft of the United States Navy and one to a Royal Navy trawler. The RCAF's ratio of kills to attacks was 7.7 per cent, fractionally higher than that of Coastal Command, and 'Undoubtedly, several Eastern Air Command attacks that were close to the mark would have resulted in serious damage or kills if the Canadian aircraft had carried the latest armament.' EAC's average of one sighting for every 134 sorties flown was only about a quarter of Coastal Command's ratio — a figure probably attributable to the smaller number of U-boats operating in the northwest Atlantic, the foul weather so often encountered by the Canadians and EAC's earlier tendency to patrol at too low an altitude.

One other maritime reconnaissance squadron needs to be mentioned here. No 413 had been formed overseas, at Stranraer on the southwest coast of Scotland, at the end of June 1941, and it was posted to Sullom Voe, in the Orkney Islands, when its Catalinas became operational in October. The commanding officer was Wing Commander R.G. Briese, RCAF, and both the flight commanders were RCAF officers although only ten per cent of the aircrew were Canadian when they went to the Orkneys, where their primary duty was covering the Home Fleet's base at Scapa Flow.

No 413's maturation was delayed by a rapid turnover of commanding officers. On 22 October an unescorted photo-reconnaissance of the Norwegian coast was ordered, a sortie 'well within the range of enemy fighters and far beyond the protection of friendly ones.' Unwilling to refuse the task (as a fellow RAF squadron commander had done), Briese included himself in the crew as a supernumerary pilot and the Catalina took off into the pre-dawn darkness — never to be seen again!

Under Briese's successor, a Canadian in the RAF, four Catalinas were sunk at their moorings in a storm, a mishap that simply should not have happened, and the CO promptly lost his job. Another RCAF officer, Wing Commander J.D. Twigg, took his place but found himself in a personality clash with the singularly immature RAF station commander. Twigg, in turn, was replaced by Wing Commander J.L. Plant, RCAF. By then, Coastal Command was probably anxious to get this problem-plagued squadron off its hands, and it was transferred to the Far East — to Ceylon (now Sri Lanka), where the first Catalina to arrive started patrolling the Indian Ocean on 3 April 1942. The very next day a second machine, piloted by Squadron Leader L.J. Birchall — seven of his eight crewmen were RAF — sighted a Japanese fleet about 550 kilometres south of the island and managed to get off a warning signal before being shot down by carrier-borne fighters. One crewman went down with the Catalina and two, badly wounded, were killed in the water by machine-gun fire from the fighters. Birchall and five others were taken prisoner when a Japanese destroyer reached the scene.

Thanks to an exaggeration by British prime minister Winston Churchill, Birchall has gone down in history as 'the saviour of Ceylon,' the implication being that his warning somehow saved the island from a Japanese invasion. In fact, the Japanese had no intention of storming ashore and were simply seeking the remnants of the British Far Eastern fleet, which had been withdrawn far out of range. However, when their aircraft subsequently launched raids on Ceylonese harbour facilities and docks, thanks to Birchall's warning the RAF was ready to meet them. That was hardly saving Ceylon; Birchall's real distinction lay in the outstanding and courageous leadership he provided in Japanese prison camps, where he saved many lives.

As for 413 Squadron, over the next three years it flew innumerable patrols from its main base at Koggala and sub-bases established along the East African coast, but its aircrews never again saw a Japanese surface vessel and managed only four attacks — all unsuccessful — on enemy submarines.

The majority of flights were, like those in any other coastal general reconnaissance squadron, long and monotonous. Whether a crew was tasked to patrol shipping lanes, do a sweep, or search a specific area, the problems were much the same. Excitement was the exception and yet maximum effort and attention were vital.

* * *

In the fighter war, Dieppe seemed to have marked the high point of German opposition in the west. The Eastern Front was now costing Hitler dearly. When not grossly outnumbered, the Luftwaffe could still dominate in the air — as yet the Red Air Force had nothing to match the Messerschmitt or Focke-Wulf and their pilots were generally inferior — but the Eastern Front stretched over 2,500 kilometres and the Luftwaffe could not be everywhere at once. When the Russians achieved local superiority they exacted a heavy toll. They may have been losing more men and machines than their enemy, but Soviet production was significantly greater and still growing.

In the west the Eighth Air Force, using heavily armed Boeing 'Flying Fortresses,' began flying daylight raids into Germany, beyond the range of contemporary fighter cover, early in 1943. In contrast to Bomber Command's loosely directed 'stream' of bombers relying on evasion by night, the Americans flew in tight formations that enabled them to bring hundreds of .5-calibre machine-guns to bear on one or more enemy fighters and endeavoured to fight their way to their objective and back. The results were disastrous, culminating in two attacks on Schweinfurt, deep in southern Germany. Sixty bombers were lost on each occasion — over nineteen per cent of the attacking force on the first raid and an incredible twenty-six per cent on the second.

Other raids, to nearer targets, proved nearly as costly but the Luftwaffe was also suffering, for the Fortresses, with their heavy weight of fire, took a toll of their own.* In the last three months of 1943 twelve German aces, who had claimed over a thousand Allied aircraft between them (143 of them in the West, but only twenty-six at night, the latter at the expense of Bomber Command), fell to the guns of the Americans.² Meanwhile, the greatest beneficiaries of this awesome struggle were the pilots of Fighter Command, who now bore little of the suffering while reaping all the advantages.

For a brief spell the Americans were compelled to pause, but their offensive resumed as the Merlin-engined North American P-51B fighter — which the British and Canadians called the Mustang — and its even-better successor, the P-51D, came into operational service. Fitted with external, droppable, auxiliary fuel tanks, the Mustang could go wherever the bombers went. By the end of 1943 the Americans could launch as many as 1,500 aircraft on any given day and, although each Fortress carried only half the bomb load of a Lancaster or Halifax, their sheer numbers meant that the enemy had to endeavour to prevent them from reaching their targets, hoping to discourage those they could not kill. But the price was great and the advent of the Mustang would raise it even further.

German day-fighter forces had mostly been withdrawn into Germany so that they could spend the maximum amount of time (and fuel) combatting the American onslaught. Fighter Command, with seven RCAF day-fighter and three fighter-reconnaissance** squadrons on strength by the summer of 1943, continued to fly sweeps over the Continent with the Germans rarely choosing to challenge them. When engagements did occur, the edge now usually lay with the Canadians.

In September 1943, for example, 402 and 416 squadrons, thirty strong, were escorting a flock of Martin Marauders assigned to bomb the railway marshalling yards at St. Pol, some eighty

---

* Which, though substantial was exaggerated to an unbelievable degree in quite absurd claims.
** Nos 400, 414 and 430 (Army Co-operation) squadrons were re-designated as Fighter-Reconnaissance squadrons at the end of June 1943, to reflect more accurately the nature of their work.

Flight Lieutenant Henry Wallace 'Wally' McLeod of Regina, the RCAF's most successful fighter pilot (21 victories over Malta and northwest Europe). He was killed in action on 27 September 1944. (CFP, PL 11993)

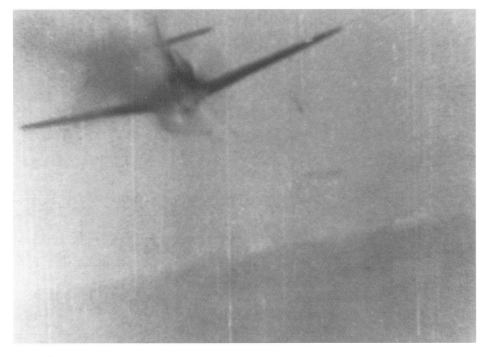

A Messerschmitt 109 under attack by Squadron Leader Albert U. Houle, No 417 Squadron, 7 February 1944. (CFP, PMR 77-520)

kilometres south of Calais. On their way home they encountered half their number of enemy fighters and the ensuing skirmish cost the Germans six machines while the Canadians lost only one — and even the pilot of that one was plucked from the sea by a motor launch of the air/sea rescue service. All the bombers got safely home.

The Canadians, like their RAF and USAAF comrades, spent much of their time strafing ground targets. By the fall of 1943, 20-mm cannon were replacing the .303-inch machine-gun on the Spitfire Vs flown by most of the RCAF fighter squadrons, and the VC's 'universal' wing enabled it to carry two 115 kg bombs on pylons under the wings in addition to its cannon.

Thus armed, the Mk Vs became fighter-bombers, better equipped to attack ground targets, while the higher-performance Mk IX Spits provided cover against the threat posed by Focke-Wulf 190s. Flak was the greatest threat that the fighter-bomber pilots faced, especially around static targets such as airfields and bridges, while railway trains — especially the locomotives — and barges on the canals and rivers that criss-crossed Holland and northeastern France were relatively easy marks.

There were four night-fighter squadrons also with Fighter Command, three of them — Nos 406, 409 and 410 — engaged in defensive operations over the United Kingdom, and one — No 418 — busy with offensive sorties over the Continent. It may be that, by and large, the fortunate few who flew these Intruder operations had the best possible war for an airman — a stimulating blend of danger and excitement but not so much of the former that it posed a threat to morale. There were occasional victims of German night-fighters and ground fire and natural hazards of weather and terrain, but casualties were never very high compared with those incurred on anti-shipping or strategic bomber operations, or even day-fighter sweeps.

Flying in DeHavilland Mosquitos, the fastest machine in general service on either side until the appearance of jet aircraft in late 1944, the Intruders had every chance to strike and evade. In September 1943, for example, No 418 accounted for eight enemy aircraft while losing two of their own, one complete two-man crew and the navigator of the second.

One other RCAF fighter squadron was not serving with Fighter Command at all. No 417 Squadron, after forming in England where it became an 'efficient Spitfire squadron,' arrived in Egypt in June 1942. It was intended that it become a 'focus' for the thousands of RCAF personnel now serving with the RAF in the Middle East, but that was a quite unreasonable expectation given the vast area involved. Indeed, it leaves one wondering if the senior officers in Ottawa who expressed such a hope could read maps or understand their scale.

Initially there were no aircraft available for No 417 and it must have seemed that they were destined to fill a purely symbolic function, but in September they were finally issued with obsolescent Hurricanes. Complaints at a political level in London put them back on Spitfires in October (just as General Sir Bernard Montgomery was winning the decisive battle of El Alamein), and they finally joined forward elements of the Desert Air Force in time to participate in the Tunisian battles that drove the Afrika Korps out of North Africa. However, the inexperienced Canadians were employed in a marginal role, largely to cover convoys moving along the coast.

Perhaps that was just as well, for the squadron was clearly not yet competent to participate in the major air battles that raged over a shrinking German perimeter. On one escort patrol it was surprised by more than twenty Me 109s, and one very lucky survivor, explaining that the Germans had come at them 'out of the sun,' went on to tell how he had reacted.

I opened fire at extreme range and closed to about 200 yards [180 metres], *using up all my ammunition in the process.... As soon as my ammunition was exhausted I took violent evasive action and discovered four 109s on my tail. After doing a steep spiral down to sea level I headed for the coast with two 109s on my tail.... Halfway across the peninsula one aircraft left me and I continued to the east coast with one 109 after me.... I shook* [off] *the last enemy aircraft and continued to base.*

Four of his compatriots were shot down.

On 10 July 1943 Allied ground forces landed in Sicily — aside from raiding operations such as Dieppe, the first Allied troops to fight on the European mainland since the evacuation of Greece in April 1941. At first from Malta, and then from Sicilian bases, the Canadians played a growing part in the air war. They were learning fast, although a month went by before Flight Lieutenant Albert Houle scored the squadron's first kill of the Italian campaign.

After that the tempo picked up. On 3 September British troops landed at the toe of the Italian 'boot' and a week later the Americans stormed ashore at Salerno, just south of Naples. The long, slow, hard-fought advance up the Italian peninsula began, with 417 Squadron now playing a full part in the process. German appearances in the air were relatively rare, however. On 30 November one such appearance enabled the Canadians to claim two enemy aircraft destroyed and one 'probable.' Between 22 January and 29 March 1944 No 417's pilots met the enemy in twenty-four engagements. They were never taken by surprise, as they had been in Tunisia, and were credited with nineteen victories while losing six machines of their own. Only one Canadian was killed, although several others were wounded.

Half a world away, on the west coast of North America, in June 1942 two other RCAF fighter squadrons, Nos 111 and 118, together with No 8 (Bomber Reconnaissance) Squadron, were engaged in combatting the Japanese, who had just occupied Attu and Kiska, two of the westernmost Aleutian islands. The Aleutians were part of Alaska and thereby technically the only part of the United States to be taken by an enemy since British and Canadian forces had occupied parts of the eastern mainland states in 1814, 128 years earlier!

Aleutian weather was probably the worst in the world from an airman's point of view, and its dangers were compounded by a lack of suitable maps covering the mountainous coastlines that defined the area. When 8 Squadron's ten Bristol Bolingbrokes moved from Vancouver to Yakutat, near the base of the Alaskan 'panhandle,' 'there were no air navigation maps of the terrain north of Prince Rupert, and the squadron made do with a few Admiralty charts as far as Juneau.' There, in primitive fashion, 'the last leg of the route had to be traced from local maps before the aircraft could fly on.'

Maps were not the only things in short supply. Spares for the Bolingbrokes were difficult to obtain and slow in arriving when they could be obtained. The Canadians relied largely on the Americans for support but their allies could not help them in this matter of spares for 'foreign' aircraft, and 'without a reliable supply system for its special parts the squadron was never able to become fully operational.' The fighter squadrons were equipped with the Curtiss P-40E Kittyhawk, however, so that spares were not a great problem for them.

Nor was the Canadians' initial prospect of actually seeing some action very great, for the Americans were understandably anxious to rely on their own units as much as possible. Wing Commander Gordon McGregor, who had distinguished himself with 401 Squadron during the Battle of Britain and now commanded the Canadians in Alaska, told his superiors that:

> ...the Canadian Squadrons will only find themselves in a location likely to result in active operations as a result of some completely unforeseen enemy attack.
>
> ...the greatest care will be taken to ensure the Canadian Squadrons will not see action if it is possible to place US Army Air Corps Squadrons in a position to participate in such action, even if the said US Squadrons are much more recent arrivals in Alaska.

Discussions at a political level subsequently won McGregor's men the right to participate in a meaningful way, and the Americans agreed that 111 Squadron should move forward to Umnak Island, the most advanced of the American air bases. Perhaps, in Casey Stengel's immortal words, the Canadians should have 'stood in bed.' On 16 July McGregor was leading the first seven machines from Cold Bay, the last air base on the mainland, towards Umiak Island when the weather began to close in and he ordered the formation to turn back. As the other aircraft followed him into a 180° turn they lost contact with him, however. Four of them, flying through thick fog, crashed into Unalaska Island, one disappeared (and has never been found) and one other continued on to Umiak. McGregor himself returned safely to Cold Bay.

When the Americans established an air base even further west, on Adak Island, the Canadians went with them, now flying American P-40Ks with long-range tanks. On 25 September, while engaged in a strafing raid over Kiska, the squadron commander, Squadron Leader K.A. Boomer, encountered and shot down a Japanese seaplane. *'I climbed to a stall practically, pulled up right under him. I just poured it into him from underneath. He flamed up and went down'* Boomer (who would be killed over Germany in 1944) had scored the only aerial kill to be recorded by a member of the Home War Establishment and became one of two RCAF pilots to be credited with victories against both the Germans and the Japanese.

Many less exciting missions were flown over the next eleven months, with the weather proving a greater threat than Japanese airmen. The Aleutian campaign culminated in May 1943 in a bloody American re-occupation of Attu and, in August, a joint American-Canadian landing on Kiska that proved to be unopposed — Attu was further west and consequently the Japanese had abandoned Kiska a few days earlier. There was nothing more to be done. The Japanese had been driven off American soil and in September 1943 the last of the RCAF squadrons returned to Canada.

* * *

In the summer of 1942 it had been agreed with the Air Ministry and the RAF that the RCAF should eventually have its own bomber group of eight or more squadrons and, since Victory Aircraft in Toronto was gearing up to produce four-engined Avro Lancasters, that the Canadian group, when it was formed, would do so on Lancasters. Sir Arthur Harris, however, was not convinced that Canada would be able to produce sufficient Lancasters to equip a squadron, never mind a group.* Therefore, when six more

---

* He was half right. Canadian production was fraught with problems and Victory Aircraft never did produce the number of Lancasters required to equip and maintain a group.

RCAF bomber squadrons were formed in the second half of 1942 (each of them starting out on twin-engined Wellingtons, as was to be expected) all except No 426 subsequently graduated to the four-engined Handley-Page Halifax during 1943. No 426 went directly from Wellingtons to Lancasters, although all but three of the others would eventually make the switch to Lancasters — some after the war in Europe had actually ended.

There was much disagreement among aircrews as to which was the better bomber, the Halifax or the Lancaster. Those who flew the former swore by it; those who flew the latter believed their machine to be the better of the two. On the surface there was little to distinguish them in terms of range, speed or ceiling, although the edge probably lay with the Lancaster, but by mid-1943 operational statistics (to which the crews were not privy) showed that only one Lancaster was lost for every 130 tonnes of bombs dropped, as against one Halifax loss for every fifty-five tonnes dropped.

Halifax losses were also proportionally greater, although only eleven per cent of Lancaster crew members survived being shot down, while the corresponding figure for Halifax crews was twenty-nine per cent. In the brutal analysis of war, far above the level of the crews being analysed, that second statistic counted for little, however. There was even a small indirect benefit in having men taken prisoner rather than killed. In either case the crews were lost to Bomber Command, but as prisoners of war those who survived being shot down became a drain on their captors' resources.

The better survivability of the Lancaster and, additionally, the fact that it was far more economical to produce in terms of labour expended per ton of bombs dropped, ensured that, as and when it could, Bomber Command would choose the Lancaster over the Halifax. For the time being, however, both were desperately needed. Aircraft were harder to come by than crews. Through the second half of 1942, Bomber Command continued to strike, mostly vainly, at German targets and absorb growing losses. But the German night-fighter force, too, was taking its lumps — as much from accidents brought about by landing and taking off in mar-

Corporal Claude Garlough and Leading Aircraftman Bill Wingrove reload 20-mm cannons on a Spitfire, No 421 Squadron. (CFP, PL 41866)

ginal conditions as by the guns of the RAF and RCAF.

In January 1943 US president Franklin Roosevelt and British prime minister Winston Churchill, together with their personal staffs and the combined American and British chiefs of staff, met at Casablanca, on the Atlantic coast of Morocco, to establish a joint strategy for the prosecution of the war. From their deliberations came an instruction to Sir Arthur Harris and General Carl Spaatz, commanding the Eighth Air Force, to engage in 'the progressive destruction and dislocation of the German military, industrial and economic system, and the undermining of the morale of the German people to a point where their capacity for armed resistance is fatally weakened.'

One of their instruments for doing this — one among many — was No 6 (RCAF) Bomber Group, formed on 1 January 1943 and initially comprising ten squadrons (three more would join it later). The Group became operational under the command of Air Vice Marshal George Brookes, formerly the air officer commanding, No 1 Training Command, in eastern Canada. The new group's bases were established in northern England, among the rugged moors of Yorkshire, not by any means the best sites for operations against most German targets. Too often, Six Group aircraft had to travel further than those of the other heavy bomber groups to reach their objectives. Moreover, northern weather left much to be desired compared with that of eastern and southern England, and the high ground that surrounded the Vale of York posed an explicit threat to aircraft — sometimes damaged aircraft — and tired pilots seeking to land after six, seven or even eight hours in the air. Flying Second World War heavy bombers was demanding work at the best of times and danger is, of itself, physically exhausting.

In part for these reasons, and also because during the first half of 1943 seven squadrons were converting from one type to another and the group as a whole was relatively inexperienced, Six Group's initial loss rates were significantly higher than those of other groups. Moreover, for the first six months the rate rose steadily — 2.8 per cent in March, 5.1 in April, 6.8 in May, 7.1 in June! In the two latter months the rate was well above the five per cent at which the psychologists had concluded that operational efficiency would begin to decline to an unacceptably low level. Six Group faltered but did not break.

Four lucky squadrons were spared the worst of this Gethsemane. The senior Canadian unit, 405 Squadron, was transferred to the élite Pathfinder Group in April 1943, and Nos 420, 424 and 425 (under the command of Group Captain C.R. Dunlap) were sent off to North Africa in May to reinforce the air assault on the Italian peninsula. Life in Tunisia was very different from that in Yorkshire, according to one of the groundcrew:

...we have become accustomed to the sun and sweat, sand and flies.... No wet canteen to go to when work is done, though. We get half a bottle of beer per week, sometimes....

Our '48s' [48-hour leave passes] *are spent at a rest camp on the Mediterranean, where we live the life of Riley. Not quite like Port Stanley* [a summer resort on Lake Erie] *perhaps, no music or pretty figures, but lovely water and cool breezes. And if you care to, you can bargain with the countless Arabs for grapes, melons and almonds, and if you are lucky a bottle of 'Vino Rouge.' Altogether not a bad life....*

For the aircrews there was a spice of danger to their lives, as opposed to the intimidating losses incurred in operations against German targets. After flying 2,182 sorties over Sicily and Italy but losing only eighteen machines on operations, the three squadrons returned to the United Kingdom by ship in early November 1943. 'I have recently seen some account of the exceptionally good work done by the Canadian Wellington Wing in the Mediterranean,' the chief of the air staff, Sir Charles Portal, wrote to Air Marshal Harold Edwards, the air-officer-commanding, RCAF Overseas.

I am told that the scale of effort in relation to the size of the force has probably been higher than has ever been achieved anywhere in the past, and included operations on 78 out of 80 successive nights with a nightly average of 69 sorties.... We are all greatly looking forward to the time when, with

Armourer, No 425 Squadron, cleans the guns of a Wellington's tail turret, North Africa, 1943. (CFP, PL 18266)

Loading rockets on a Beaufighter, No 404 Squadron. (CFP, PL 41009)

newer and better equipment, they will resume their operations against Germany.

Senior officers such as Portal and Edwards may have been 'looking forward to the time' with enthusiasm, but it is unlikely that the crews Portal was praising felt quite the same way about returning to the strategic bomber offensive.

During 331 Wing's sojourn in North Africa, the rest of the Group had taken part in the first great firestorm raid when Sir Arthur Harris launched nearly 800 bombers — seventy-eight of them from Six Group — against Hamburg on the night of 27/28 July 1943. Because they were sheltered by *Window* — aluminium foil strips that spread widely as they floated slowly down, forming a screen impenetrable to the enemy radars — which had just been cleared for use, losses were low, while the increase in numbers of four-engined bombers carrying heavier bomb loads meant that a greater weight of bombs had been dropped than in the three 'thousand-bomber' raids to date. That, combined with high temperature, low humidity and a juxtaposition of two unusual weather fronts, created an inferno on the ground below.

> ...concentrated bombing produced a firestorm which covered as much as five square miles [thirteen square kilometres] of the city centre. Large and ever-growing fires raised the temperature at the core to several hundred degrees and this superheated air rose so rapidly that it sucked in behind it great quantities of cooler, oxygen-rich air at velocities approaching hurricane strength.... These winds encouraged fires on the periphery, spreading the conflagration further — all while the bombing continued. Firefighting was impossible in such circumstances and even those flying far above the city were soon aware that something extraordinary was happening. Canadian crews...were all emphatic that Hamburg was blazing.... The smoke from the fires was so thick that it penetrated into the cabins of the bombers, almost choking the crews.... Hamburg was blazing like a paper box.

The official tally of those killed in the raid was 41,800 but the true figure was probably much higher. In many of the city's cellars a heap of ashes with a few charred bones was the only evidence of death and no one could say how many corpses it represented.*

The rare weather conditions required to create firestorms ensured that they could not be made to order, however, no matter how much Sir Arthur might desire them. In October a smaller but still awful firestorm developed at Kassel, while the third and last of the European war did not happen until the bombing of Dresden in February 1945.

Sheer unadulterated courage counted more than ever as Bomber Command vainly tried to win the war by itself and took terrible casualties in the process; but, paradoxically perhaps, night operations were increasingly reliant (on both sides) on scientists — 'boffins', the English-speaking airmen called them — working mostly in the still-esoteric realm of electronics. British boffins strove to enable bomber crews to navigate more accurately, while their German peers worked on ways to locate bombers in the cloud-riven night skies.

The Germans devised *Freya* and *Giant Würzburg* radars to detect the bomber stream before it could reach its targets, and the bomber crews answered with *Window* and with *Mandrel* jamming, at first transmitted from England but subsequently airborne as well; the enemy responded with *Freya Fahrstuhl*, *Mammut* and *Wassermann* radars with anti-jamming capabilities.

The bombers used directional *Gee* and *Oboe* radio beams projected from England to help locate their targets; the defenders countered with *Heinrich* and *Bumerang* jammers. The attackers turned to *H2S* — downward-looking radar that provided a picture of the terrain below, for the bomber to navigate by — and the enemy matched that with *Naxos* and *Naxburg* homing devices, which allowed enemy crews to locate and attack bombers using *H2S*. Night-fighters used their *Lichtenstein* airborne interception radars to search out bombers, and their

---

* German air raids on Britain killed about 50,000 over the whole war.

No 402 Squadron fighter pilots pass the time while waiting for a call to scramble. (CFP, PL 22672)

Pilots and navigators of No 418 Squadron at pre-flight preparations before embarking on Intruder sorties over Europe. (CFP, PL 40805)

opponents introduced their own AI radar, *Monica*, which could pick up night-fighters sneaking up from behind; but the Germans produced a counter-measure for *Monica*, so that it, too, facilitated interception. Radioed instructions coming from ground-controllers were jammed or confused by *Fidget*, *Drumstick* and *Jostle* transmissions from England and, later, airborne *Tinsel*. On one raid, for example, 'on many fighter control frequencies little could be heard except a bizarre collage of Hitler's speeches and readings of the poetry of Johann Wolfgang Goethe.'

All these exotic tools and techniques came into play incrementally, spread over the length and breadth of the bomber offensive, each with its own grotesque identifier — *Cigar* and *Airborne Cigar*, *Corona*, *Dartboard*, *Grocer*, *Piperack*, *Perfectos*, *Shiver* and *Spanner* on one side, *Donnerkell*, *Dudelsack*, *Erstling*, *Flamme*, *Laubfrosch*, *Lux* and *Sägebock* on the other. There were more, for as each new device or modification was introduced it was frustrated by another. There seemed no end to it, but in front of this pivotal science, above and below and all about it, was the awesome courage of the bomber crews and the night-fighter pilots who took to the air night after night, adjusting knobs, refining signals and reading meters and gauges until they could drop their bombs or fire their guns to best effect. The bomber offensive of 1942–43 and the first months of 1944 was the Second World War's equivalent of the First World War's Somme and Passchendaele.

The cycle of raids on primarily industrial targets continued unabated, but another exceptional attack was that on Peenemünde, sufficiently important to earn its own code-name. Operation HYDRA was directed against the German rocket-development complex there, its existence known through painstaking intelligence work. HYDRA, if it was to be successful, demanded precise accuracy rather than Bomber Command's usual style of area bombing, and to achieve that Harris assigned a senior Pathfinder officer to act as 'master bomber,' circling the target (in a Mosquito) throughout the raid and providing:

> the bomber force with minute to minute information regarding the progress of a raid, issue warnings of misplaced markers, give the position of dummies and generally assist the bomber force in successfully attacking the correct aiming point. It is further hoped that such commentaries will serve to strengthen the determination of less experienced crews....

The RCAF's Pathfinder unit, 405 Squadron, contributed twelve aircraft to HYDRA and the squadron commander, Wing Commander John Fauquier, was one of the master bomber's two deputies.* Six Group provided another sixty-two aircraft. Extensive damage was done and Germany's V-2 rocket programme was set back by two months or more, but the cost was high. In bright moonlight, numbers of night-fighters got into the bomber stream and forty crews were lost — almost seven per cent of those dispatched. Six Group, in the last wave of bombers, suffered worst of all, with twelve machines — half of them manned by veteran crews — failing to return, for a loss rate of nearly twenty per cent.

In the past the German capital, Berlin, had been a sporadic objective of Bomber Command despite its distance from British bases (in the region of 850 kms), the fact that it was beyond the range of *Gee* and *Oboe* — which were restricted to 'line of sight' between a bomber and the transmitter in England — and the need to fly over a considerable stretch of enemy territory to reach the city, thereby giving the Luftwaffe more time to react. However, in the fall of 1943, as winter approached and the nights got appreciably longer, Sir Arthur Harris turned his attention towards Berlin. The battle of Berlin began on 18/19 November with 440 Lancasters, including twenty-nine from 408 and 426 squadrons, assigned to attack the city in the brief span of sixteen minutes. Simultaneously, another force of nearly 400 bombers, including a hundred Halifaxes of Six Group, attacked Ludwigshafen, 500 kilometres to the south.

---

* In subsequent raids Fauquier would often act as Master Bomber, and eventually he would command the RAF's crack bomber unit, 617 Squadron, made famous by the 'Dambusters' raid of 16/17 May 1943. He ended the war with a DSO and two Bars and a DFC, having flown three tours of bomber operations.

Over Berlin, where the weather was abysmal and the cloud cover heavy, only nine bombers were lost, none of them from the Canadian squadrons; over Ludwigshafen, where it was clear, twenty-three failed to return, including seven from Six Group. Both raids were assessed as failures, the Germans recording only seventy-five high-explosive bombs falling on Berlin, while the bombing of Ludwigshafen was reported as 'widely scattered.'

Four nights later more than 750 aircraft went to Berlin for a raid scheduled to last no more than twenty-two minutes. This time the bombing was concentrated and accurate — over 1,750 Berliners were killed, nearly 7,000 injured and 180,000 left homeless, with the Telefunken, Blaupunkt, Siemens and Daimler-Benz factories all severely damaged. Twenty-six bombers were shot down, the Canadians losing only two crews, both from the RCAF's 'hard luck' squadron, No 434.*

Other targets were selected for the rest of the month but on 2/3 December another purely Lancaster assault was launched against the German capital. This one was a disaster, costing Bomber Command a total of forty aircraft, but Six Group was lucky again, this time losing only two out of thirty-five machines dispatched. Was it a case of improved training and/or a matter of seasoning? The experience of one 424 Squadron crew gives us some indication:

While flying over the target, their aircraft was coned by 50–70 searchlights.... The M[id] U[pper] [G]unner first sighted a Ju 88 on the port quarter down at 400 yards [375 metres]...and gave combat manoeuvre corkscrew port and the fighter immediately broke off his attack. No exchange of fire.... The second attack developed from starboard quarter down and the MUG...gave combat manoeuvre starboard and again fighter immediately discontinued his attack and broke off.... No exchange of fire. The third attack came from the port quarter down. Again MUG gave combat manoeuvre port and the enemy broke off....

Fourth attack developed from starboard quarter down...and MUG once again gave combat manoeuvre...and again fighter immediately discontinued his attack.... The fifth and last attack developed from port quarter down.... Enemy aircraft came in to 60 yards [55 metres] range and broke away to port beam above, giving MUG sitting target...sparks and tracer were seen to ricochet off fighter and enemy aircraft dived steeply.... The rear gunner was completely blinded throughout these five attacks by the blue master...and other searchlights.

It was probably the same fighter which attempted all five attacks, and the enemy pilot's inability to follow the unwieldy bomber in its corkscrew turns suggests that he was not very good at his job. To that extent, the Canadians were lucky. But much credit must go to the mid-upper gunner and pilot of the bomber for the speed and vehemence of their reactions, which were certainly not those of inadequately trained novices.

On 16/17 December Berlin was struck again, this time by nearly 500 bombers in really miserable weather. More than a hundred large fires were started and over 500 civilians killed but only 'a handful' of industrial facilities were destroyed. Twenty-five machines were lost to enemy action, but worse was to come: low cloud and fog covered most of Britain and thirty-four aircraft, including three from No 405, Canada's Pathfinder squadron, crashed on landing. Total losses were fifty-nine, twelve per cent of the crews dispatched.

Improving weather enabled high-flying Mosquitos to photograph the damage inflicted to date.

> An examination of the statistical analysis of damage shows that over 1,250 net acres [500 hectares] of business and residential property has been affected in the fully built-up...areas covered by these sorties and that over 60 per cent of the buildings in the Tiergarten district alone have been destroyed. Very substantial figures are also given for Charlottenburg, Mitte, Schöneberg, Wedding, Wilmersdorf and Reinckendorf.

Exceptionally foul weather in Yorkshire kept Six Group out of the next Berlin raid, on 23/24 December, but four nights later the Group con-

---

* No one has been able to explain why 434 Squadron's losses should have been consistently higher than those of other squadrons in the Group; but they were.

tributed 143 sorties to a 712-strong main force — their largest effort to date. Overall losses were only 2.8 per cent but the Canadians lost 3.5 per cent of their crews.

The New Year brought more attacks on Berlin, interspersed with attacks on a variety of industrial targets across the whole width and depth of Germany. Six Group participated in most of them and continued to take their share of losses, but no longer more than their share, through January and February of 1944 when the Battle of Berlin, at least, ended for the time being. Bomber Command had other fish to fry, as we shall see in the next chapter.

The final accounting on Berlin was inconclusive. Much damage had been inflicted on both sides but, as was usually the case, the RAF/RCAF attackers were better able to deal with it than the Luftwaffe and the German economy. In neither case had the damage been decisive.

The overall loss rate suffered by Bomber Command over the winter of 1943–44 — leavened by raids on easier targets — was fractionally above the magical five per cent figure that psychologists believed was the maximum that could be tolerated without a significant loss of operational efficiency. Five per cent losses meant that only twenty out of a hundred crews could expect to survive the standard tour of thirty sorties. In Six Group, which had endured significantly higher losses than the other groups before the Battle of Berlin began, the rate had begun to drop — a trend that would only accelerate under Air Vice-Marshal C.M. McEwen, 'a demon for standards and training,' who took over command of the Group from George Brookes in February 1944.

Chapter VI

MASTERS OF THE AIR, 1943–1945

Prior to May 1943, rather more than two thirds through the war against Germany, the RCAF's anti-submarine efforts had accounted for only three U-boats, in whole or in part; between that date and the end of the war, two years later, the toll increased by another sixteen, for a total of nineteen, four of them in the remainder of 1943 and twelve in 1944–45.

In January 1943, 407 Squadron, which had suffered so severely in the course of anti-shipping strikes during 1941 and 1942, was converted to Wellington aircraft equipped with Leigh lights and re-designated a 'general reconnaissance' squadron, that term being a euphemism for anti-submarine duties. Leigh lights were very powerful floodlights that could be used to illuminate the sea surface at night — when U-boats were most likely to be on the surface re-charging their batteries — after centimetric radar had brought an aircraft into the immediate vicinity. Thus equipped, No 407 flew its first anti-submarine sortie in March 1943 and claimed its first victory in September 1943, with three more to follow over the next fifteen months.

Squadron nos 422 and 423 were formed as general reconnaissance squadrons, and were soon flying Sunderland flying-boats over the North Atlantic from bases on Lough Erne, in the western tip of County Fermanagh, Northern Ireland. Both squadrons spent most of their air time acting as distant guardians to convoys, and their sorties were usually long and often boring. But action, when it did come, could be exciting to say the least. By mid-1943 most U-boats carried a substantial anti-aircraft armament and were not reluctant to use it.

The two squadrons claimed three U-boats between them, in addition to one kill shared with two destroyers (one of them a Canadian vessel, HMCS *Drumheller*) in May 1943. No 423 Squadron's first exclusive success came in August 1943, when Flying Officer A.A. Bishop sank the enormous (1,700-tonne) re-supply submarine *U-489* in the Western Approaches. Caught on the surface, the enemy was well armed, with an automatic 37-mm gun and a battery of quadruple 20-mm cannon, and made no attempt to submerge, choosing instead to fight it out and no doubt expecting to win if the airmen were so foolish as to press home an attack in such a slow-moving and unwieldy machine.

We turned in towards the U-boat at around 1,200 yards [1,100 metres] *and opened fire with our .5-inch* [12-mm] *machine-gun...then we opened fire with our .303 Vickers* [machine-gun].

At this point the Jerries who, as far as we could tell, hadn't hit us yet, started to register a few. From there on in it was a steady rain of lead, wounding the second pilot and the second wireless operator who was down in the nose on the [.303 machine-] *gun.*

We managed to hang on and dropped our six depth charges right up the track of the U-boat from dead astern.... By the time we had released the depth charges the aircraft had a terrific fire in the port wing, the aileron controls and elevator trimming tabs were shot away and things didn't look so good....[1]

The Sunderland crash-landed and six of the twelve crewmen, including a severely wounded wireless operator, saved by Bishop, scrambled into an inflatable dinghy before it sank. Meanwhile the U-boat was also sinking, and its crew abandoned ship.

The Jerries sat quite comfortably on their rafts...and made no attempt to come and pick us up. The first wireless operator saw smoke on the horizon, but none of the rest of us knew anything about a destroyer coming until it was right beside us and had launched a whaler to pick us up. They had seen us go down to attack, and followed the smoke from our burning fuel to our position.

No 423 Squadron scored another victory in October, and its accomplishments, together with the series of successes enjoyed against surfaced U-boats by other anti-submarine squadrons and naval vessels over the summer and fall of 1943, led to the introduction of Schnorkel 'breathing' tubes by the Germans in early 1944. Schnorkels enabled submarines to run on their diesel engines while submerged to periscope depth, so that it was feasible, if hardly practicable, not to surface at all throughout a cruise. That made finding them from the air much harder. While the RCAF squadrons in Coastal Command sank, or shared in the sinking of, nine U-boats in 1944–45, only two such successes were achieved after the D-Day landings.

Finally, there was one RCAF anti-submarine squadron working the mid-North Atlantic that was not part of Coastal Command. Formed in May 1942 on Consolidated Cansos, the Canadian variant of the Catalina, No 162 first spent an uneventful eighteen months on the east coast as part of the Home War Establishment, before being lent to Coastal Command in January 1944 and transferred to Iceland. From there, 162 Squadron accounted for its first U-boat, 700 kilometres southwest of Reykjavik, in April 1944.

Then a lull in North Atlantic operations (while the long-range U-boats were being fitted with schnorkel) sent the squadron from Reykjavik on temporary assignment to Wick, on the northeastern tip of Scotland. There, in June 1944, the month that brought D-Day and the invasion of northwest Europe, No 162 enjoyed an extraordinary run of successes — and losses! The first triumph came on 3 June, just three days before the Normandy assault, when a squadron Canso caught *U-477* on the surface no more than 400 kilometres south of the Arctic Circle. Perhaps the U-boat commander felt that it was safe to linger on the surface so far north. If so, he was fatally mistaken; attacking through heavy flak, Flight Lieutenant R.E. McBride and his crew dropped a well-placed salvo of depth charges. The *'U-boat appeared to lift bodily and swing to port, losing almost all forward movement, then submerged on an even keel.... At least five survivors were seen in the water,'* but even in midsummer the sub-arctic water was too cold for anyone to survive for any length of time.

A week later, Flying Officer L. Sherman's Canso caught another U-boat on the surface only a little further south, and sank it with the loss of all hands. But a similar fate was about to befall Sherman and all but one of his crew. In the early hours of 13 June their Canso was so badly damaged by flak in an unsuccessful attack on *U-480* that it crashed into the Norwegian Sea. Five of the crew of eight — not including Sherman himself, who must have drowned —

Servicing a Lancaster II. (CFP, PL 26009)

The bomb load of a Halifax bomber. (CFP, PL 29078)

managed to get into an inflatable dinghy and then drifted without food or water for nine days, four of the five dying. Only hours from death himself, Flight Sergeant J.E. Roberts, who had resisted the temptation to drink seawater, was rescued by a Norwegian fishing boat. He was nursed back to health in a German hospital and dispatched to a prison camp.

On the same day that Sherman was shot down, another No 162 Canso, commanded by Flying Officer J.M. McRae (with the squadron commander, Wing Commander C.G.W. Chapman, on board as second pilot), sighted a periscope track in the same general area where the squadron had met its earlier successes. The submarine surfaced and opened fire just as the Canso attacked, and both flak and depth charges were successful. The U-boat sank, and after putting out an SOS call the flying-boat crash-landed and sank as well. The Canadians had to take to a small dinghy, two climbing into it and the other six simply holding on. Two and a half hours later another aircraft dropped them a lifeboat but by then one of those in the water had already died of hypothermia. The seas were too high for a flying-boat to pick them up, and an air/sea rescue launch took six hours to reach them, too late to save two more of the crew. The five survivors were each decorated — the second pilot getting a membership in the Distinguished Service Order while the aircraft commander had to be content with a Distinguished Flying Cross!

The last of this notable string of victories came on 24 June, when a Canso under the command of Flight Lieutenant David Hornell sighted and sank *U-1225* in the still chilly summertime water some 200 kilometres north of the Shetland Isles. As he bored in to the attack, flak from the submarine set the starboard wing on fire and when the engine fell out of the wing Hornell had to crash-land his machine in the sea. It was still burning as it sank and there was only one dinghy for eight men. Like McRae's crew, ten days earlier, they had to take turns in it, two at a time.

In this case, the wireless operators had not succeeded in putting out an SOS call, the radio being damaged by enemy fire, but luckily another aircraft spotted them after they had been in the cold, sub-arctic water for five hours. Once again the sea state prevented a flying-boat pick-up and when a rescue launch arrived, sixteen hours later, two crewmen had already died and Hornell was past saving. He was posthumously awarded the Victoria Cross for 'pressing home a skilful and successful attack against fierce opposition with his aircraft in a precarious condition, and by fortifying and encouraging his comrades in the subsequent ordeal.'

Following on the formation of 413 Squadron, whose story was recounted in the last chapter, No 415 had been formed as a torpedo-bomber squadron, flying obsolescent Bristol Beauforts, in August 1941. The squadron never flew Beauforts operationally, however, soon converting to Handley-Page Hampdens — which were very little better!

No 415 was a 'hard luck' squadron if there ever was one, not so much in terms of losses as in every other respect. Training on the Hampdens was made more difficult by frequent shifts of base through the spring, summer and fall of 1942 — a month in Cornwall, three weeks in Hampshire, two months in Lincolnshire, a month at Wick.... Not until November 1942 did the unit return to Hampshire — to Thorney Island — for a year-long stint, at which time it was re-trained on *two* different types of aircraft simultaneously, half the squadron on a torpedo-carrying version of the Wellington and the other half on Fairey Albacores.

The Albacore was a three-seater, single-engined *biplane* with a *maximum* speed of 260 kph and a pitiable defensive armament of one fixed forward-firing machine-gun and two flexibly mounted guns in the rear. Other, luckier, torpedo-bombing squadrons flew heavily-armed, twin-engined Bristol Beaufighter Mk XCs with a maximum speed twice that of the Albacore. Indeed production of this miserable machine ceased just about the time that 415 Squadron began to get them.

Given the sorry aircraft that the squadron had to work with and the risks inherent in slow, low-level strikes (torpedoes demanded that the attack be made slowly from a minimal height, lest the torpedo break up on impact with the water) through often intense flak, perhaps it was just as well that there was also an on-going

The sinking of U-625 by No 422 Squadron, 10 March 1944. (CFP, RE 68586)

Lancaster bomber raid on Haine St. Pierre marshalling yards, 8/9 May 1944. (CFP, PA 202795)

shortage of torpedoes which slowed training even further.

The squadron claimed its first so-called success at the end of May 1942, off the French coast, when three Hampdens attacked the largest ship in a six-vessel convoy escorted by five flak ships. The merchantman was claimed as 'damaged,' but as one Hampden failed to return it seems questionable whether the affair could fairly be assessed as a success.

Subsequently, when the squadron was split into Wellington and Albacore flights, it was split physically as well, the Wellingtons flying out of two bases in Norfolk (one of which also housed squadron headquarters) while the Albacores were based at Manston, in Kent, and Thorney Island. By this time the Albacores were no longer carrying torpedoes — rather, they were laden with 40-kg anti-submarine bombs, which they were supposed to drop on E-boats, the manoeuvrable, high-speed launches that the Germans used for night-time raiding in the English Channel and southern North Sea. It was an impossible assignment. The E-boats, when they could be found, were just too nimble.

As for the Wellingtons, they ended up carrying no air-to-surface weapons at all. Instead, they were required to locate, report on and shadow E-boats bent on mayhem by night or convoys creeping down the Norwegian, Danish or German coasts under cover of darkness. When anti-shipping strike forces — aircraft or ships — reached the vicinity, they were expected to drop flares and illuminate the enemy. It was one-sided, frustrating, unfulfilling work.

Morale sank from bad to worse until, in February 1944, Air Marshal Lloyd Breadner, the former chief of air staff who had become Overseas AOC-in-C on 1 January, suggested to Ottawa that 'you demand its [415 Squadron's] withdrawal from Coastal Command where it has never had a decent role. It could be allocated to either Bomber Command or Tactical Air Force.' However, it took time for such a radical change to be put into effect and during that time the squadron achieved its one indubitable success. In March 1944 one of the Wellingtons homed four torpedo-armed Beaufighters onto a convoy near Borkum, off the German coast. In the light of Canadian-dropped flares, the 'Torbeaus' sank one ship without loss. And finally, at the end of July 1944, the change took place and No 415 Squadron became Six Group's fourteenth — and last — bomber squadron, flying Handley-Page Halifaxes.

The RCAF's anti-shipping strike squadron, No 404, had a much happier time flying Beaufighters from September 1942 until the last month of the war. The Beaufighter was a versatile aircraft, and theirs were armed with rocket projectiles which enabled them to use all their speed in attack. The rockets could be fitted with different types of warhead, depending on whether the user was assigned to flak suppression (high-explosive, one being the equivalent of a 15.5-cm shell) or the actual sinking of small ships (armour-piercing). Usually they were assigned to suppress or sink the small flak escorts, leaving the larger freighters — many of them carrying vital iron ore with a high phosphorus content from Swedish mines — to their torpedo-armed brethren.

No 404 moved south to assist in covering the flanks of the invasion force on D-Day, and was part of a massive strike force which encountered three German destroyers sailing towards the invasion front on the evening of 6 June. In the initial attack, this time with RAF Beaufighters providing the anti-flak protection, the Canadians hit two of the destroyers with armour-piercing rocket projectiles and left one on fire. Early the next morning they were back to claim a number of further hits. None of the three was sunk or immobilized but they were all compelled to seek shelter in Brest and, when they left port again on the 9th, a destroyer flotilla that included HMCS *Haida* and *Huron* sank one, drove a second ashore and forced the third to return to Brest once more.

A steady diet of success followed, in both the Channel and the North Sea, culminating in a massive sweep into the Baltic on 4 May 1945 — just three days before the unconditional surrender of the German forces in the west and the end of the war in Europe. On that occasion, forty-one Mosquitos, including seven from 404 Squadron which had just converted from the Beaufighter, attacked a convoy of *two* freighters and *five* escorts — the proportions suggest how desperate the Germans had become — sinking one of the freighters and badly damaging the other.

* * *

The Allied fighter forces had won their war some time before that. By the spring of 1944 practically all the German fighters in the west — which were none too many, even before that — had been withdrawn from France and the Low Countries in response to the US Eighth Air Force's Mustang-escorted attacks on German targets. The experience of Flight Lieutenant W.R. McRae, on his third operational tour in the year preceding D-Day, illustrates the unequivocal effect of that withdrawal.

On June 1st, 1944, I was...completing my first year on operations with 401 Squadron, but had yet to fire my guns in air-to-air combat.... The thought of returning to Canada having to admit that, not only had I failed to shoot anything down, but that I had not even shot at an enemy aircraft, was haunting me. I had engaged in ground attacks and dive bombing of course, but...I believed that a fighter pilot's role was to destroy enemy aircraft....²

Even on D-Day itself the Luftwaffe could do little to counter the overwhelming mass of Allied aircraft supporting the invasion. Only about 300 aircraft of all types were available to meet the 11,000 that the Allies deployed on 6 June 1944. Two squadrons among dozens, No 442 sighted a pair of Focke-Wulf 190s that wisely fled the scene before they could be attacked and No 401 tried to intercept a formation of about twenty enemy machines that also turned away before the Canadians could engage them. *'While the Squadron was climbing to ward off this danger, a lone enemy aircraft sneaked through...dropped a single bomb on the beach and made good its escape,'* in one of the only two instances where German fighter-bombers succeeded in attacking the beachhead that first day.

There was of course plenty to do by way of strafing the enemy. Indeed in November 1943 six Home War Establishment fighter squadrons had been transferred to the United Kingdom and renumbered in the 400-block to play their parts in the air-to-ground war. Three of them, Nos 438, 439 and 440, had been equipped with Hawker Typhoon fighter-bombers — known colloquially as 'Bombphoons' — capable of carrying two 800-kg bombs. And once they had dropped their bombs their great speed and four 20-mm cannon made them formidable low-level fighters.

Their performance fell off alarmingly at altitude, but there were many squadrons of Spitfire IXs and XIVs to deal with any German pilot bold enough to try and interfere from above. Among them were the other three RCAF squadrons from the Home War Establishment, now renumbered 441, 442 and 443 and equipped with Spitfire IXs, but so slight was the quantity and quality of German opposition that they also frequently indulged in strafing of ground targets. Altogether there were, during the northwest Europe campaign, ten Canadian day-fighter squadrons in the field, organized into three wings, Nos 126, 127 and 144,* each of which indulged in much more air-to-ground strafing than air superiority combat, together with three fighter-bomber squadrons (143 Wing) and three fighter-reconnaissance squadrons (39 Wing).

(Still off by itself, far away in northern Italy, No 417 Squadron, flying Spitfire VIIIs from August 1943, was mostly engaged in dive bombing — something the Spit was singularly unsuitable for, given its streamlined form — and air-to-ground strafing in the near-total absence of the Luftwaffe.)

Let one outstanding day speak for all. On 18 August 1944 the German forces in western France were trying to escape encirclement in the Falaise 'pocket' and all of the Second Tactical Air Force was concentrated on stopping them. No 127 Wing reported 'the busiest day in its history, claiming to have destroyed or damaged almost five hundred vehicles in 290 hours of flying time, expending about 30,000 rounds of 20-mm ammunition in the process.' *'A number of our aircraft were hit by flak and several crash landed away from base but only one pilot, F/O Leyland of 421 Squadron, went missing.'*

---

* No 441 Squadron flew with 125 (RAF) Wing before returning to England in October 1944.

On the ground:

The acrid smell of burning and burnt-out vehicles was bad but the stomach was turned by the stench of the dead men and horses — and there were thousands of dead horses. The smell was all-pervading and overpowering. So strong in fact that pilots of light artillery observation aircraft flying over the area reported that the stench affected them even hundreds of feet in the air.

Above the battlefield shimmered a miasma of decay and putrefaction...and the unburied Germans, swollen to elephantine grossness by the hot sun...lay with blackened faces in grotesque positions....

Once in a long while the Luftwaffe fought back, but now most of its pilots were ill-trained and inexperienced. Perhaps that explains how Flight Lieutenant Richard Audet of 411 Squadron came to shoot down three F-W 190s and two Me 109s in a single sortie on 29 December 1944. In fifty-two earlier sorties flown since joining the squadron in mid-September, he had fired his guns only three times and those without success as his opponents left the scene at top speed. However, he would register five more victories — one of them against one of the new jet fighters, the Messerschmitt 262 — before he himself was killed by flak while strafing a railway yard on 3 March 1945.

Despite such occasional outings as had enabled Audet to shoot down five in one day, the Luftwaffe continued to fade, making only one last, futile attempt to inflict serious damage on the Allied fighter forces. Gathering together every possible machine and a lot of half-trained pilots, at first light on New Year's Day 1945 a task force of some one thousand aircraft was launched against eleven of the most important Allied airfields on the Continent, hoping to catch all the Allied pilots on the ground, many of them perhaps suffering hangovers.

At Eindhoven, which housed the RCAF's Mustang fighter-recce and Typhoon fighter-bomber wings as well as several RAF squadrons, the Germans were relatively successful. This was one of several occasions during the later stages of the northwest Europe campaign when non-flying personnel came under enemy fire.* The station medical officer reported:

...in a very few minutes the floors of our large ward and of the corridors were literally covered with a mass of gaping wounds and bloody uniforms. One airman in the corner was screaming and waving in the air two mangled hands that hardly seemed attached at all to the arms.... One chap, who had had trouble with varicose veins, looked at his shattered leg and said, 'They sure fixed my veins for me.' He died the next day. Another, whose arm was almost severed at the shoulder, was saying, 'If they put me to sleep and I wake up without my arm, I shall go crazy.'[3]

Thirty-one aircraft were destroyed on the ground; thirteen airmen were killed and dozens wounded (five Canadian non-fliers were among those killed and fifteen were wounded) for the loss of sixteen Luftwaffe machines and their pilots. But the RAF and RCAF losses, in men and machines, were easily replaced: German losses were not.

At Heetsch the Germans caught ten Spitfires of 401 Squadron just lining up for take-off, but they all managed to get into the air and then shot down six of the enemy without loss to themselves. RAF fighters, already in the air and informed by radio, intercepted the attackers and claimed another thirteen for the loss of one of their own and another pilot wounded. At Evère, the third Canadian base, the Germans accounted for twenty-four of sixty Spitfires on the ground but lost eleven of their own to flak.

Overall, the Allied forces lost 160 aircraft destroyed and half that number damaged, but the enemy suffered even more, losing 300 machines and 214 pilots — a third of them to their own flak, since the German anti-aircraft gunners along their route were unaccustomed to seeing any but Allied aircraft and had not been warned of the impending gamble. But even had the casualty totals been reversed, or the Allies suffered twice or thrice as heavily as

---

* Another such incident arose on 8 April 1945 when the Typhoon Wing was ordered forward to an airfield near Osnabruck, in Germany. The new airfield was still garrisoned by a few German soldiers, who used 'mortars and 88-mm guns... After sustaining some casualties our men moved in on the 11th, though the occasional Hun still crept out of his foxhole with the cry of "Kamerad"' — *The R.C.A.F. Overseas: The Sixth Year* (Toronto: University of Toronto Press, 1949), 299.

Armourers Leading Aircraftman Jack Peachell (Winnipeg) and Corporal Ozzie Shaw (Windsor) check the 20-mm cannons used by No 409 Squadron's night-fighting Mosquitos. (CFP, PL 31818)

Halifax bomber of No 420 Squadron over Le Havre, September 1944. (CFP, PL 32846)

Spitfire IX flown by Wing Commander Lloyd V. Chadburn. One of the privileges of his rank was to have his aircraft marked with his own initials. (Canadian Forces photograph, RE 641811)

they actually did, it would have made little difference. In twenty-four hours their losses had all been replaced; the Luftwaffe's losses could never be made good.

German scientists had long been working on so-called secret weapons connected with the air war, mostly defensive in nature, the first significant offensive one being the V-1 'buzzbomb' or 'doodlebug.' This was a small, unmanned, ramjet machine laden with a ton of high-explosive that continued on a pre-set course until it ran out of fuel. It was, obviously, extraordinarily inaccurate.

Air intelligence had revealed the forthcoming V-1 campaign and efforts to destroy the camouflaged and well-protected launching sites had been in operation for the past eight months, slowing but not entirely thwarting their operational debut. Had the enemy been in a position to start a full-scale offensive on the first day, London might have been as devastated as Cologne and Hamburg already were and surely a substantial part of the fighter force covering the invasion would have had to be withdrawn to counter it; but happily the V-1 only trickled into service.

The first V-1, fired from an earthen ramp in the Pas de Calais, fell on England on 12 June 1944 — just six days after Allied forces landed in Normandy. Only the fastest fighters could hope to shoot down buzzbombs, which were capable of reaching speeds in excess of 700 kph, but anti-aircraft fire, concentrated in a belt south of London, took a heavy toll, as did a greatly reinforced balloon barrage over the city.

The first one to fall to RCAF guns was claimed on the night of 16/17 June by a Mosquito crew of 418 (Intruder) Squadron, assigned to patrol over the Channel and keep a look-out for these miserable instruments. At least they were easy to see with their conspicuous reddish-orange exhausts — so easy that the squadron accounted for three that first night.

We used to stooge around just out from the launching area in France. We were the first line fighter patrol. Sometimes we could see the actual launchings.... Then, when the thing came up and it could be spotted by the steady glow from the rear end, we dived down vertically at full throttle.... After our dive we would level out and let go with a quick burst. Then, if you were too close you'd be thrown all over the sky by the explosion, or flying debris would damage your machine.[4*]

By mid-September 1944, when more than a thousand V-1s had been shot down by fighters (RCAF crews accounting for almost a tenth of the total), that part of the French coast from which the V-1s were launched was overrun by First Canadian Army and the threat was past. The V-2, which came too late in the war to have a decisive impact, was a long-range rocket which travelled in a stratospheric arc at previously unheard-of heights and speeds. There was of course no riposte to the V-2 except for the defeat of Germany, but the first of them did not strike London until September and, in total, only about a thousand fell on English soil — half of them on London — causing less than half the casualties of the V-1 attacks.

Mention was made earlier of 'light artillery observation aircraft' over the Falaise pocket. Artillery observation, in the guise of 'army co-operation,' had been the primary purpose of 400 Squadron when it and its Westland Lysanders had arrived in England in February 1940; but both 400 and 414 squadrons, formed for the same purpose in August 1941, had subsequently been re-equipped with Curtiss Tomahawks and then with North American Mustangs, to be converted to a 'fighter-reconnaissance' role in early 1942.

Since then the wheel had turned full circle and, with total Allied domination of the skies over every Allied battlefield, the idea of — and need for — artillery observation had come back into favour. In June 1944 General H.D.G. Crerar, the commander of First Canadian Army, had recommended the formation of new army co-operation squadrons — now to be labelled Air Observation Post squadrons, since they would have no other function but artillery observation. The first of them, No 664 (because their function would be entirely with the army they were not numbered in the 400 block), was formed in England in December 1944. No 665 Squadron

---

* Eventually, pilots overtaking a V-1 in a dive learned to nudge its wing with their own, causing it to tilt off course and crash into the sea or rural areas.

Pilot Officer Andrew Mynarski, VC. (CFP, PL 38261)

Flying Officer Bernard C. Denomy and Flight Lieutenant David Hornell. Denomy was awarded a DSO for his role in the action, while Hornell earned a posthumous VC. (CFP, PL 30823)

was added a month later. Both were equipped with Austers.

In some respects the Auster was not unlike the Lysander; both were slow-moving, single-engined, three-seater, high-wing monoplanes, but the Auster was more agile, could idle along *in the air* at 65 kph and had a shorter take-off and landing capability. It carried no armament but if attacked had a good prospect of surviving a fighter attack provided it stayed *really* close to the ground. More than one Auster pilot escaped by finding a copse or even a single, large bushy tree and flying around it in tight circles until his higher-performance enemy, unable to close to a killing range, gave up in disgust and went home.

No 664 flew the first of its 619 missions on 29 March and No 665 flew the last of its fifty-eight on 7 May, the former losing one aircraft and two aircrew killed and the latter losing none of either in the course of their short war.

* * *

The bomber offensive reached a crescendo in 1944–45, with the Eighth Air Force bombing by day and Bomber Command bombing by night — a schedule extended to day and night once bases had been established in Holland and fighters could escort the bombers to most objectives.

In the spring of 1944 the Berlin excursions had become simply too expensive for Bomber Command to continue at the frequency endured over the previous winter and Sir Arthur Harris reverted to emphasizing closer targets that required less time spent in German air space, thereby reducing loss rates to more manageable proportions. That was a particular relief to Six Group, whose casualties to date had been so heavy that it had been quite unable to build up the cadre of seasoned crews needed to ensure that lessons learned were properly propagated. In a vicious circle of waste, the more crews that were lost, the greater the likelihood of more becoming lost.

There was considerable variation in loss rates month by month, but between March 1943 and February 1944 Six Group's Halifax loss rate never dropped below 4.5 per cent and only very briefly (at the end of July) below five per cent — which, the reader will recall, was supposed by the experts to be the maximum rate at which morale could be maintained. In January 1944 losses reached an unbearable 9.8 per cent and the Halifax crews briefly faltered. Nor were things much better in the Lancaster squadrons — just over seven per cent in December and just under seven per cent in February; January was something of a frolic at only 5.5 per cent. At that point their morale was marginal, too.

Little mention has been made in these pages of decorations awarded to RCAF airmen for gallantry. There were a great many of them, every one well deserved, but with the notable exception of two Victoria Crosses it would surely be iniquitous to remark on some but not others and quite impracticable to mention all. It is the authors' belief that every man who flew in Bomber Command, decorated or not, deserved a medal for valour. A 9.8 per cent loss rate, as with the Halifaxes of Six Group in January 1944, extended over a tour of thirty operations, reduced the chance of survival to a fraction more than one in twenty! It was a virtual death warrant.

The question of whether losses would have been significantly fewer if there had never been an RCAF bomber group — if the Canadian squadrons had continued to be distributed among RAF groups, as they had been before the formation of Six Group — was debated then and has been much debated since. It seems likely that they would have been lower but an element of Canadian sovereignty would certainly have been lost.

However, relief was at hand. On 1 April Bomber Command was placed under the ultimate direction of American General Dwight D. Eisenhower, in preparation for the seaborne assault on northwest Europe. While on occasion his crews might continue to attack industrial centres in Germany, Sir Arthur Harris was now required to devote the bulk of his resources to hitting transportation targets — bridges over the Rhine and various French rivers and railway yards and junctions throughout Belgium and northern France — that would make it difficult for the enemy to reinforce his ground forces when the Normandy invasion began in June.

A Dakota with RCAF ground crews arrives in Normandy as RCAF fighter squadrons take possession of dusty airfields built under fire by the Royal Engineers. (CFP, PL 30046)

Canso 'A' of No 162 Squadron — the machine piloted by Flight Lieutenant David Hornell during his VC exploit. (CFP, PMR 77-147)

A Typhoon of No 143 Wing taxies out for another bombing mission. An airman on the wing guides the pilot, who cannot see over the nose. (CFP, PL 42813)

This shift to much closer objectives, combined with the intensive training imposed on even operational units by its new commander, Air Vice-Marshal McEwen, made a particularly dramatic impact on Six Group's loss rates. By April the rate for Halifaxes was down to two per cent, never again to rise above 2.5 per cent; and for Lancasters it was down to under one per cent and would only fractionally exceed three per cent in July. Indeed during the last year of the European war Six Group's loss rates would be the lowest in all of Bomber Command.

But statistics are one thing, personal experience is another and for those crews shot down it often made little difference whether their fates overtook them in French or German skies. Against French targets, misfortune most often struck in the form of flak but there were still fighters to deal with on occasion. On the night of 12/13 June 1944 a Lancaster of 419 Squadron returning from a raid on the Cambrai rail yards was 'attacked from below and astern by an enemy fighter:'

> As an immediate result of the attack, both port engines failed. Fire broke out between the mid-upper turret and the rear turret, as well as in the port wing. The flames soon became fierce and the captain ordered the crew to abandon the aircraft.
>
> Pilot Officer [Andrew] Mynarski [the mid-upper gunner] left his turret and went towards the escape hatch. He then saw that the rear gunner was still in his turret and apparently unable to leave it.... Without hesitation, Pilot Officer Mynarski made his way through the flames in an endeavour to reach the rear turret and release the gunner.... Eventually the rear gunner clearly indicated to him that there was nothing more he could do and that he should try and save his own life. Pilot Officer Mynarski reluctantly went back through the flames to the escape hatch...and saluted before he jumped out of the aircraft.... Both his parachute and his clothing were on fire. He was found eventually by the French, but he was so severely burnt that he died from his injuries.

Ironically the rear gunner, Flying Officer G.P. Brophy, was flung clear when the Lancaster hit the ground moments after Mynarski jumped and his turret broke away from the burning fuselage. Recovered and concealed by the French Resistance, Brophy was back in England by September to tell his story — a tale of valour that brought Mynarski a posthumous Victoria Cross, the second and last awarded to an RCAF airman.*

Later in August the RCAF got its third George Cross** when a Halifax of 425 Squadron, landing at Tholthorpe on three engines after an attack on a V-1 site in the Pas de Calais, clipped a parked aircraft which was fully loaded with bombs and set it on fire. As shards of exploding bombs whistled past them, the base commander, Air Commodore A.D. Ross, and a number of airmen struggled to release the pilot and rear gunner trapped in the returning plane. Ross led the rescue, personally wielding an axe until a bomb fragment 'practically severed' his arm between the wrist and elbow; 'he calmly walked to the ambulance and an emergency amputation was performed on arrival at station sick quarters.'

In another notable sacrifice virtually identical to David Hornell's VC exploit, Flying Officer R.D. Gray, the navigator of a Wellington serving in 172 (RAF) Squadron, won the RCAF's fourth and last George Cross at the end of August 1944. Gray's pilot took his machine in to attack a U-boat in the Bay of Biscay but the aircraft was shot down by a hail of flak. Only one small dinghy inflated, too small to hold the four crewmen who survived the flak and crash. Despite

---

* Squadron Leader Ian Bazalgette, RAF, who was born in Alberta but emigrated to England at the age of nine, won a VC in the course of a bombing raid on 4 August 1944. Lieutenant R.H. Gray, RCNVR, a British Columbian, was awarded a posthumous VC while flying with the Royal Navy's Fleet Air Arm against the Japanese on 9 August 1945.

** The first two Canadian GCs were won posthumously by Leading Aircraftman K.M. Gravelle on 10 November 1941 and LAC K.G. Spooner on 14 May 1943. Gravelle lost his life attempting to save the pilot of a crashed Tiger Moth; Spooner lost his when he took control of an Avro Anson when the pilot lost consciousness and held it in the air while three fellow student navigators baled out successfully. The Anson then crashed into Lake Erie, taking Spooner and the pilot to their deaths.

A Mosquito XII night-fighter with bulbous radar nose. (CFP, PL 46955)

Lancaster X at Yarmouth, Nova Scotia, 13 June 1945, as bombers and aircrews returned to Canada in preparation for 'Tiger Force' and the war against Japan. (CFP, PMR 98-159)

Liberators of No 11 (Bomber Reconnaissance) Squadron and Lancasters returned from No 6 (RCAF) Group, Bomber Command, lined up at Yarmouth, Nova Scotia, June 1945. (CFP, PMR 98-158)

being wounded himself, Gray helped two of his injured comrades into the dinghy and then clung on until, overcome by exhaustion, he let go and drowned. The two men in the dinghy were picked up by a warship the next day.

Nearly fifty-five years after the event, one wonders why Hornell should have received a Victoria Cross and Gray a George Cross. Two nearly identical situations, each resulting from direct involvement with the enemy. The VC, then and now, seems to carry slightly more prestige in the public eye, although the two awards acknowledged similar resolution and dedication.

When Bomber Command reverted to the air staff's control in September 1944, Sir Arthur Harris once again began to put all his effort into attacks on German targets. But now the Allied armies were mustered along the German border and the distances to be flown over hostile territory were proportionately less, and the Luftwaffe was not the threat it had once been. The shift in emphasis was accompanied by the introduction of daylight raids — an experience that affected some crews in an unexpected way. In one of the earliest daylight raids, they 'got no satisfaction out of their Emden attack despite or because of the fact that they could see what was happening down below,' reported RCAF historian Flight Lieutenant F.H.C. Reinke, gathering data for a future history. 'They couldn't help thinking about the people down there. The centre of the town was the aiming point.'

In October Bomber Command launched the largest of all its raids when 1,055 aircraft were dispatched to Essen, in the Ruhr. Unlike the thousand-bomber raids of 1942, this time no crews from training units had to be included in order to put so many machines in the air. Moreover, all those that participated were four-engined 'heavies,' so that a far greater weight of bombs was delivered. Bombing through cloud, the attackers caused 'extensive damage' to a complex of Krupp factories but lost only twelve crews — three (0.65 per cent) of them from Six Group. Thirty-six hours later a daylight raid brought 771 raiders back to the same target. Essen, like many other German cities, was now little more than a heap of rubble.

By February 1945 'from the Ruhr to the Swiss frontier,' crews were told:

> ...there is no sizeable town...which has not suffered serious damage at the hands of the Allied air forces with the single exception of Wiesbaden.... It is proposed to let Wiesbaden share the state of most other German towns and by destroying it to eliminate one of the last few remaining places where the German army can be assured of sound shelter from the rigours of winter.

On the night of 2/3 February that, too, was done.

Perhaps a similar justification might be offered for the destruction of Dresden, which was brought about by 800 bombers on the night of 13/14 February — including sixty-seven Lancasters from Six Group and ten from the Pathfinder Group's 405 Squadron. However, 'that it was really a military necessity...few will believe,' admitted Harris's deputy, Air Marshal Sir Robert Saunders, long after the war had ended.[5] The bombing created a firestorm and probably about 50,000 civilians died, with another 50,000 being burnt or otherwise injured.

As civilian casualty figures rose, air force loss rates dropped, although there were still nights when Six Group suffered severely. On 5/6 March 1945, 185 Canadian machines participated in a raid on Chemnitz. Nine crashed immediately after take-off, the result of icing, another six were lost to flak, while fighters over the target and a further three crashed on landing. That meant a ten per cent loss rate and 104 dead airmen. The figure was terrible in itself, but it was nothing when compared to the damage being done to Germany. At the end of the month Winston Churchill told his chiefs of staff that:

> the moment has come when the question of bombing German cities simply for the sake of increasing the terror, although under other pretexts, should be reviewed. Otherwise, we shall come into control of an utterly ruined land.... The destruction of Dresden remains a serious query against the conduct of Allied bombing. I am of the opinion that military objectives must henceforward

be more strictly studied in our own interest rather than that of the enemy.

A month later, on 25 April 1945, Six Group flew its last sorties, against coastal batteries on Wangerooge in the Frisian Islands, and lost five aircraft — twenty-eight aircrew — in a chain reaction of collisions that began when one Lancaster, caught in the slipstream of another, slipped sideways into a third.

With the end of the European war, Six Group aircraft were engaged in ferrying freed prisoners-of-war back from Germany to the United Kingdom, while at the end of May the eight squadrons which were to form the RCAF component of 'Tiger Force' in the forthcoming assault on the Japanese home islands — Nos 405, 408, 419, 420, 425, 428, 431 and 434 — began to fly their Lancasters back to Canada. However, the atomic bombing of Japan brought that war, too, to a timely end on 15 August 1945.

The statistics of the bomber offensive are complicated. Many RCAF aircrew flew in RAF squadrons, some of the RCAF squadrons flew in RAF groups before the creation of Six Group and some non-Canadian Commonwealth airmen flew in RCAF squadrons before and after the formation of Six Group. Then there was 405 Squadron, which, for a good part of its Second World War epic, was part of the Pathfinder Group. All we can say with certainty is that 9,919 RCAF airmen lost their lives in Bomber Command and that Six Group's contribution to that gallant band was 4,272 — the vast majority, but not all, being Canadians.

* * *

One other aspect of airpower — air transport — came to have a Canadian component during the last nine months of the war. Nos 435 and 436 squadrons were formed in India, in November and August 1944, respectively, and No 437 was formed in England in September 1944. All three squadrons flew Douglas Dakotas.

No 437 was the first to fly operationally, and its introduction to battle was a fiery one. Ten aircraft, laden with equipment and towing Horsa gliders containing infantrymen, were a part of the aerial armada that carried the British 1st Airborne Division to Arnhem on 17 September 1944, as part of the ill-fated Operation MARKET GARDEN. There were no losses on that first day, the Germans having been caught by surprise, nor on the next three days despite an increasing amount of flak. But on the 21st five of ten crews were lost, and dropping the supplies where they were needed was easier said than done.

The cold-blooded pluck and heroism of the pilots was quite incredible. They came in in their lumbering machines at fifteen hundred feet searching for our position.... The German gunners were firing at point-blank range and the supply planes were more or less sitting targets.... How those pilots could have gone into it with their eyes open is beyond my imagination.... They came along in their unarmed, slow, twin-engined Dakotas as regular as clockwork. The greatest tragedy of all, I think, is that hardly any of these supplies reached us.[6]

Subsequently, 437 Squadron flew a variety of missions, transporting urgently needed supplies and reinforcements to the Continent, bringing casualties back to England and carrying mail both ways. One of the squadron's more unusual assignments came on 24 December, when it delivered to Antwerp and Melsbroek some 20,000 lbs [over 9,000 kgs] of Christmas puddings destined for troops bogged down in the mud along the Dutch-German border. In its eight months of wartime flying, No 437 lost fourteen aircraft and seventeen aircrew.

The two transport squadrons in southeast Asia were kept busy servicing the British Fourteenth Army, which was fighting the Japanese in Burma. Their first sorties were flown in December and January 1945, but their war went on longer than that of 437 Squadron, for the Japanese did not surrender until August. They also had unpredictable weather (and the monsoon) to contend with. Nevertheless, between them they moved 56,000 tonnes of supplies and 30,000 passengers in eight months, losing eight aircraft, fourteen aircrew and six passengers in the process.

* * *

In round figures, some 13,000 RCAF personnel were killed on operations or died while prisoners-of-war during the Second World War and another 4,000 died in training accidents or from other causes. Thirty of them were members of the Women's Division. Nearly 1,555 were wounded or injured and 2,475 became prisoners-of-war. In the summer of 1945, as the killing stopped, the RCAF was the world's fourth-largest air force and perhaps the best of them all, but the price had not been small. 'If blood be the price of admiralty,' Rudyard Kipling had written in his *Barrack Room Ballads* half a century earlier, 'Lord God, we ha' paid in full!'

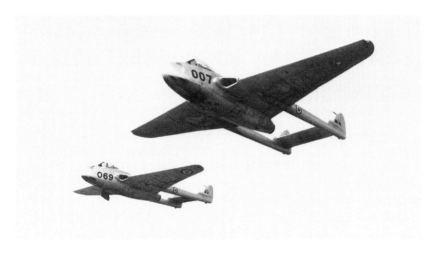

Chapter VII

FROM WAR TO PEACE TO COLD WAR

With the end of the Second World War came many tasks. Once the men had been welcomed home, the parades staged and thousands discharged to 'civvy street,' there remained the matter of those who had not survived. Every former war theatre had its complement of unmarked graves and unidentified remains. The task of identifying as many as possible fell to the Missing Research and Enquiry Services — British units with RCAF representation seeking the remains and trying to assign names to the dead. Squadron Leader William Mair, the senior RCAF officer so employed, devised the exhumation process and some of the means used to identify corpses. It was horribly unpleasant work. Eventually the teams disbanded with many casualties still missing or their bodies unidentified. It was then that construction began on impressive monuments to those who had no known grave; in the case of the Commonwealth air forces these included the memorials at Runnymede, Malta, El Alamein, Singapore and Ottawa unveiled between 1953 and 1959, on which were inscribed more than 45,000 names including those of 4,400 members of the RCAF.[1]

Sadly, in its haste to 'downsize', the air force ignored the task of writing its own history. Colonel C.P. Stacey had written of how he overcame ministerial indifference to win a paper war that left him with the resources to begin work on the Army's official histories. His RCAF counterpart, Squadron Leader (later Wing Commander) F.H. Hitchins, lacked Stacey's networks and ruthlessness. The office of Air Historian sputtered along for years with one or two officers and no clerical staff to catalogue the thousands of documents that had been shipped to his office by domestic and overseas units. Hitchins's successor, Wing Commander R.V. Manning, who assumed the post of Air Historian in 1960, had no better luck. It was not ministerial apathy that stood in the way but the myopia of the Air Council in general and successive Air Members for Personnel in particular. In spite of the fact that Stacey's work had demonstrated the value of an analytical history and despite having greater resources (the RCAF by then consumed fifty per cent of armed forces budgets), Manning's superiors at Air Force Headquarters provided nothing of substance towards the preparation of a similar history. This had to wait until the 1980s, when the first volumes were published under the aegis of the uniformed Canadian Armed Forces.

Meanwhile the RCAF war machine was dismantled through rapid demobilization. By the end of 1947 its strength stood at 12,200 all ranks, including 650 in the auxiliary. Much of the wartime infrastructure and equipment was dispersed through the Crown Assets Disposal Board. Schools were closed, then sold off. Everything had to go — the lumber from dismantled hangars, the urinals from barrack washrooms. The Royal Canadian Flying Clubs Association bought 150 Tiger Moths for $23,100; entrepreneurs in Canada and abroad closed similar deals on airworthy craft that could be converted to a commercial mode — Hudsons, Lysanders, Ansons and more. Hundreds of derelict aircraft ended up as children's playthings on farms or behind service stations. Over the next twenty years most were either torn apart or burned; the component most likely to survive would be a tail wheel (handy for a wheelbarrow). Belatedly, air museums as far away as Finland and South Africa began looking for Second World War airframes and 1945's junk became desirable artifacts.

The demobilization process was marked by a major tragedy. On 15 September 1946 a Dakota transport of No 124 (Communications) Squadron crashed and burned at Estevan, Saskatchewan, killing all twenty-one aboard. Most of the passengers were RCAF pilots detailed to fly surplus aircraft from Estevan to the United States. It was the second-worst single-aircraft accident in the history of the RCAF.*

In the immediate post-war period, the RCAF was commanded by two very different men.

From 1944 to 1947 Air Marshal Robert Leckie served as Chief of the Air Staff. An architect of the wartime British Commonwealth Air Training Plan, Leckie presided over the post-war demobilization (which he thought excessive) and the initial joint Canadian-American defence planning. Although he enjoyed friendships with many senior American officers, forged in part during fishing expeditions, Leckie disagreed with them over the extent of the Soviet threat, which increasingly dominated defence policies.

Leckie was succeeded by Wilfred A. Curtis, whose good fortune it was to direct the RCAF's initial Cold War expansion; at his retirement in 1953 it had grown to 46,100 all ranks. A genial figure with many high-level contacts in the RAF and USAF, Curtis had been infuriated by the difficulties encountered by the wartime RCAF in securing the aircraft it most needed from our Allies. As Chief of the Air Staff (and backed by both Brooke Claxton, minister of national defence, and C.D. Howe, minister of trade and commerce), Curtis urged that Canada develop, as far as possible, her own aircraft industry. This would benefit many firms, such as Canadair in Montreal and DeHavilland of Canada in Toronto, and both the rise and fall of Avro Canada between 1948 and 1959 would be rooted in policies and decisions that Curtis had fostered.

Until the establishment of the North Atlantic Treaty Organization (NATO) and Canada's commitment to collective security in Europe, much of Canadian defence policy was centred upon the north. This culminated in the concept of a Mobile Striking Force — an army brigade group, to be moved and supplied by aircraft — capable of dislodging enemy troops who might land in northern Canada. The first post-war application of the theory was Exercise MUSKOX (February to May 1946), in which a convoy of tracked army vehicles struggled to make it from Churchill to Edmonton via the Arctic shoreline and the Mackenzie River, supplied by Norseman, Dakotas and CG-4 gliders. RCAF documents described MUSKOX as a triumph; the airmen smugly observed that aircraft alone could have transported the troops and vehicles to their destinations, eliminating the need for all that driving! Army assessments were less enthu-

---

* The sad distinction of being the RCAF's worst accident was the crash of a Liberator bomber of No 10 (BR) Squadron on 19 October 1943. The airplane had been carrying personnel proceeding on leave from Newfoundland to Montreal; twenty-four persons lost their lives.

Anson aircraft being burned just after the Second World War. (Photograph provided by Larry Milberry and Canav Books, PMR 98-162)

A derelict Fairey Battle, Macdonald, Manitoba, 1963; it shows vestiges of wartime stripes that distinguished gunnery training aircraft. (CFP, RE 22019)

The graves of the Lancaster crew killed during a supply drop at Alert, Ellesmere Island, July 1950. (CFP, PA 202907)

siastic, but the groundwork had been done for larger joint exercises to come.

RCAF and army co-operation continued at many levels. In April 1947 a new formation, No 11 Tactical Air Group, assumed responsibility for planning and training in air support for land forces. It was raised to Command status in June 1953. Over the years it supervised numerous regular and auxiliary units in western Canada, including the RCAF's Survival Training School and (until 1954) the Joint Air Training Centre at Rivers, Manitoba. Joint army/air force manoeuvres were held from the Yukon to Churchill, as aircraft dropped troops, supplied them and attacked phantom opponents. By 1958, however, the army had concluded it needed air transport only; aerial tactical support as hitherto practised was of diminishing importance. Tactical Air Command was disbanded on 1 January 1959, its components being divided between Training Command and Air Transport Command.

Meanwhile the Arctic beckoned in other ways. In 1946 the RCAF launched a shoestring operation (one Canso, two Norseman) to seek airfield sites in the Arctic archipelago for future survey or rescue operations. Banks Island and vicinity was deemed particularly important. Among those involved was Wing Commander A.H. Warner; as a mechanic he had participated in the RCAF's Hudson Straits Expedition of 1927–28.

This investigation was followed by Operation POLCO (July–September 1947), aimed at fixing the location of the North Magnetic Pole. This had been the subject of increasing interest on the part of American, British and Canadian authorities, enhanced by the growing potential value (commercial and strategic) of Arctic air routes. No 413 Squadron dispatched a single Canso carrying twelve men (including four civilian specialists) to take magnetic readings at nine locations.

Overall, POLCO was both pioneering in its nature and immensely successful in its outcome. It marked the first time in Canadian aviation history that a flying-boat had been flown and based for an extended period amongst the treacherous and barren islands surrounding the North Magnetic Pole. Its findings were not as conclusive as had been hoped; two follow-up MAGNETIC operations were needed, in 1948 and 1949.

The interwar aerial photography and mapping programme had been suspended during the Second World War except where it would have a direct military bearing (i.e., airfield construction). Planning began in the spring of 1945 to resume this work, using both old methods and new techniques developed overseas. Despite an ambitious programme, however, smoke from forest fires curtailed the work. In fiscal year 1945–46 only 694,000 square kilometres were photographed, although in the next year roughly 909,000 square kilometres were covered. These gains were made by phasing out older aircraft (Avro Ansons) and substituting newer machines (North American Mitchells, Avro Lancasters) capable of flying higher and devouring more territory in their cameras.

The 1947 flying season was even more impressive. Two squadrons were tasked with conducting aerial surveys and completed over 1,024,000 square kilometres of photography. Still better results were achieved in 1948, when 2,332,000 square kilometres were photographed, the highlight coming in July when a Lancaster crew discovered two previously uncharted islands and thus added 13,000 square kilometres to the Dominion. One of these islands was named Air Force Island, reflecting the means of its discovery. As a result of the 1948 operations only fifteen per cent of the Dominion remained to be photographed for the compilation of aeronautical charts extending to seventy-five degrees north latitude. Gap-filling continued until, by 1957, the task was complete. No 408 Squadron, the principal unit in this programme, then concentrated on Arctic reconnaissance work, related more to Cold War tensions than domestic requirements.

Aerial mapping was, in large measure, a seasonal affair. Each spring and summer photographic detachments fanned out across the north to remote, even primitive airfields. Unpredictable weather might delay operations; long Arctic summer days helped. The crews followed precise tracks, photographing from 22,000 feet as they went. They were assisted in their work by SHORAN (Short Range Aid to

C-119 Flying Boxcar dropping army supplies on exercises. (CFP, PL 101513)

P2V-7 Neptune overflies a submarine. (CFP, PL 105305)

Mustang firing rockets, an example of Second World War technology persisting into the early 1950s. (CFP, PL 54765)

Navigation), which had grown out of wartime bombing technology. Portable SHORAN stations were established at increasingly remote sites. They emitted radar pulses that enabled personnel to establish precise distances between various points and thus draw the flight lines. Millions of photographs were subsequently collected at the RCAF's Photo Establishment, the 'White House' at Rockcliffe, where they were assembled into mosaics that eventually became maps.

It was a complex programme and although much of the technology was state of the art some of it smacked of improvisation. The Lancasters, which came to symbolize the programme, had been designed for neither northern nor photographic operations and needed extensive modifications, which slowed their introduction into service. Tires wore out quickly on gravelled Arctic airstrips. The Merlin engines had to be overhauled every 300 hours, yet it was not uncommon for a Lancaster to spend upwards of 400 hours per summer on detached operations. That meant frequent engine changes, not in the comfort of a Rockcliffe hangar but in the open at remote sites — Cambridge Bay, Fort Chimo, Frobisher Bay, Prince George, Fort Nelson, Yellowknife, Norman Wells and Whitehorse, to name only the most common — where flies and mosquitoes attacked everyone, except on those days when it was raining.

Formal honours were rare in the post-war RCAF, but two officers associated with the mapping work received the Trans-Canada Trophy (better known as the McKee Trophy). Flight Lieutenant (later Brigadier General) Keith R. Greenaway was given the 1952 award for his expertise in Arctic flying, meteorology and contributions to devices assisting in high-latitude aerial navigation; and Wing Commander J.G. Showler, who commanded No 408 Squadron for much of its most intense work (1954–57), received the Trans-Canada Trophy in 1958.

* * *

The Second World War had revolutionized every aspect of aviation, including Search and Rescue (SAR). Search aircraft increased their range; the airborne lifeboat was introduced; radar that had hunted submarines in any weather could now locate crippled ships at sea. Helicopters were employed to effect rescues from remote areas and dangerous terrain. Equally important were the wartime developments in survival kit, rations and clothing that gave crash victims or storm-lashed mariners a fighting chance at survival until rescuers arrived. Before and during the war SAR successes had been due as much to good luck as good management. From 1947 onwards, with the RCAF assuming international SAR responsibilities under International Civil Aviation Organization agreements, rescue missions became routine. Most were carried out smoothly and uneventfully. A few became dramatic epics.

Late in September 1947 an Anglican missionary, Canon J.H. Turner, was reported to have been seriously injured in a hunting accident at Moffet Inlet, Baffin Island, 450 miles inside the Arctic Circle. A Dakota of No 112 Transport Flight, Winnipeg, was assigned to the rescue operation. Flight Lieutenant Robert C. Race piloted the aircraft to Moffet Inlet; on 4 October, in the face of low clouds and rugged terrain, he located a frozen lake where he dropped four para-rescue soldiers led by Captain Lionel G. d'Artois. Race kept the ground party supplied until a suitable landing site could be found and marked, all the while treating Canon Turner. This entailed seven weeks of work in harsh weather, limited daylight and brutal topography. Finally an evacuation flight was arranged. Race touched down on ice untested except by ground parties and subsequently took off by the light of an improvised flare path. Canon Turner was flown to Winnipeg, 1,700 miles away, but unfortunately he later succumbed to his injuries.

Few of the RCAF's many SAR operations attracted much public attention, but one such sortie had a tragic outcome. On 21 August 1949 a Canso carrying Inuit polio victims from Chesterfield Inlet to Winnipeg via Churchill crashed in fog and rain near Norway House, Manitoba. All twenty persons aboard were killed — seven RCAF aircrew, seven patients, a nurse, a newspaper reporter and four other civilians.

The RCAF was also engaged in aid to the civil power, including major emergencies. When

'Swords into ploughshares' — a Lancaster on photo survey operations in northern Canada, 1948. (CFP, PL 48094)

North Star transport unloads supplies at Isachsen weather station, Ellef Ringnes Island. (CFP, PL 54050)

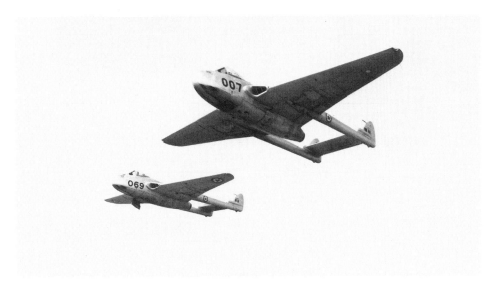

Vampire scramble. (CFP, PL 95091)

British Columbia's Fraser Valley suffered disastrous flooding in May 1948 the minister of national defence ordered the forces to assist. Flying 900 sorties, RCAF aircraft carried out evacuations, delivered over two million sandbags and carried fifty-five tons of freight (from water pumps to blood plasma) as well as several hundred soldiers. Some 450 members of the RCAF worked on the ground to complement the contribution of about one hundred aircrews. In surveying the operation, the Department of National Defence proudly observed that 'the forces had never before been called upon to give help on such a large scale in a civil crisis and their swift response and able co-operation pointed to a readiness to meet more critical emergencies should the need arise'[2]

In the spring of 1950 a similar disaster occurred in Manitoba's Red River valley. Once again the forces were deployed on a scale unprecedented for peacetime. Operation REDRAMP required RCAF transport aircraft to carry in men and relief supplies from as far away as Los Angeles. Over 500 tons of cargo were flown in — in one single forty-eight-hour period 300,000 sandbags weighing 200 tons were delivered.

* * *

Canada's major role in the Korean War was an army brigade; the RCAF contribution was secondary though significant. As of June 1950, No 426 Squadron was the only long-range transport unit in the force; it had been busy earlier that year with Operation REDRAMP. On 20 July the unit was ordered to prepare six North Stars with crews and technicians for transpacific operations. The official announcement came on the 21st; on the 23rd the unit diarist noted that Dorval and Lachine, Quebec, had been inundated with 'a plague of newspaper reporters, radio announcers, etc.' The first six North Stars departed Dorval two days later, carrying double crews, 185 ground crewmen and twenty-five tons of supplies. They flew in formation to Ottawa where they passed over the Peace Tower, dipping their wings in salute to the body of former prime minister William Lyon Mackenzie King, then lying in state in the Parliament buildings. The North Stars continued in formation to Toronto, whence they proceeded independently.

No 426 Squadron established its west coast base at McChord Field in the state of Washington, the hub of the USAF's Military Air Transport Service (MATS). Three North Stars departed Tacoma, Washington, on 27 July, arriving at Haneda (Tokyo) two days later; flying time was just under twenty-eight hours one way. For a time it was hoped that the unit could dispatch one sortie per day, but this proved beyond the resources of a six-plane outfit. Eventually 426 found it could send off five aircraft per week. On 3 December 1950 the squadron reported its one hundredth departure for Japan.

The varying fortunes of United Nations forces affected No 426 Squadron's expectations. In October, with U.S. General Douglas MacArthur's armies apparently winning, personnel had speculated that Operation HAWK might soon be ended. On 9 November, however, the unit diarist recorded 'reversals on the war front' and in December there was conjecture that 'serious reverses' would compel 426 to move its forward base to Japan in anticipation of massive airlift evacuation from Korea itself. Fortunately the situation never became so grave. When the front stabilized in June 1951 airlift requirements declined. However, No 426 continued to be involved in other tasks, such as Arctic resupply, mercy flights and transatlantic transport.

For most of Operation HAWK the North Stars flew a northern Pacific route — McChord, Elmendorf, Shemya and Haneda. Alternative fields were used at Adak and Missawa. A few trips were made via the central and south Pacific, but these encountered stronger headwinds. Early in the operation, when trained personnel were in short supply, some crews flew 155 hours a month, almost always above cloud. MATS recognized this to be excessive and set a limit of 110 hours a month. On 25 December 1951, while on a transpacific run, Flying Officer R.M. Edwards had crossed the International Date Line and gone from 24 December to 26 December. Newspapers picked up on the item, describing Edwards as 'the man who had no Christmas.'

When HAWK was wound up on 9 June 1954, the RCAF announced that 426 Squadron had

Flight lieutenants Claude Lafrance and Larry Spurr each receive the American Distinguished Flying Cross for services in Korea. (CFP, RE 21084)

flown 599 round trips (four of them directly to Korea itself), logged 34,000 flying hours, carried 13,000 personnel and airlifted 3,500 tons of freight. HAWK was conducted with no fatalities but there were some very close calls. On 15 April 1951 a North Star was cleared by Ashiya tower to descend from 4,000 to 3,000 feet prior to landing. At 3,400 feet the aircraft hit trees atop a hill that was not even indicated on the airfield map. No one was injured but the aircraft sustained considerable damage to its nose, oil cooler and exterior radio aerials. On the night of 27 December 1953 another North Star crashed at Shemya, Alaska, on the homeward run. The aircraft was landing in a fifty-knot cross wind and blowing snow. Runway ice reduced braking action and high gusts blew the North Star off the runway into a gulley.

The RCAF was not heavily engaged in the air war over Korea. As of 1950, the force had no fighters more advanced than obsolescent DeHavilland Vampires. A programme to re-equip with North American F-86 Sabres built under licence by Canadair had been announced in April 1949. The Toronto *Telegram*, moved by either extravagant anglophilia or envy of Montreal, had protested, arguing that the RCAF could be more effective and get the same value if it bought 800 Vampires rather than 130 Sabres. Other Toronto papers briefly took up the cry. Fortunately they were ignored. There would be no more procurement mistakes comparable to the dreadful Fairey Battle purchases of fifteen years earlier.

Once the RCAF began training Sabre pilots (and Canadian industry began producing Sabres for the force), the bulk of these pilots and machines were directed to home defence or to meet NATO commitments. Nevertheless, twenty-six pilots were attached to USAF Sabre units for experience in case our NATO squadrons became involved in a European war. One of these pilots, Flight Lieutenant J.A.O. Levesque, shot down a MIG-15 on 30 March 1951, becoming the first Canadian to destroy an aircraft in all-jet combat.

When Levesque returned to Canada there were no plans to inject RCAF replacements into the Korean air war. Meanwhile, American Sabre strength remained at one wing until 1952 when production of Canadian-built aircraft provided sufficient reserves for the USAF to dispatch two more wings to the Far East. The work by Canadair was this nation's most significant contribution to the battles waged in 'MIG Alley'.

Assignment of RCAF pilots to Sabre squadrons in Korea resumed in 1952. They flew limited, fifty-mission tours, as compared with the normal 100 missions for USAF pilots. That meant that just as the men were beginning to get the hang of jet combat they were pulled back to Canada. While many wanted to go to Korea, an informal 'old boys' network' sent very few young pilots; most of those posted were Second World War veterans, some with very distinguished careers. A relative tyro, Flying Officer J.C.A. Lafrance, complained to his superiors about this situation, arguing that junior pilots like himself most needed combat experience. Lafrance won his point — and a Korean assignment. He proved to be a very hot pilot; in August 1952, flying his first assignment as element leader, he destroyed a MIG-15 and was subsequently given command of a flight. Lafrance later attained the rank of major-general in the Canadian Armed Forces and became highly respected among tactical fighter leaders.

The 'old sweats' included Flight Lieutenant Ernest A. Glover, who reached Korea in June 1952. Up until 26 August he never saw a MIG; from then until the end of September he saw them nearly every day. In all, Glover was credited with shooting down three MIG-15s and damaging a further two, for which he received a Distinguished Flying Cross; he is the only member of the RCAF to win this award when Canada was formally at peace.*

In November 1952, Squadron Leader Andrew R. Mackenzie was assigned to the USAF's 39th Squadron. He flew four uneventful missions but on 5 December 1952 was accidentally shot down by another Sabre pilot. He baled out and was captured immediately. At first he was treated well but soon the enemy turned on him. Mackenzie was to endure countless interroga-

---

* Captain Peter Tees, Royal Canadian Artillery, was also awarded a DFC, for services with No 1903 Independent Air Observation Post Flight, the artillery eyes of the Commonwealth Division.

Aero-engine technicians install a new Orenda engine in a Sabre, No 4 Fighter Wing, Baden-Soellingen. (CFP, PL 83354)

Sabres of No 411 Auxiliary Squadron over Toronto. (CFP, PMR 98-169)

The very cold Cold War: Russian aircraft on a northern ice floe, photographed by No 408 Squadron. (CFP, PMR 98-163)

tions, brainwashing attempts and 465 days of solitary confinement and would eventually lose 30 kilograms. He was not allowed to write to his family until April 1954 and he was not released until December 1954 — two years after his capture and seventeen months after the Korean Armistice had been signed.

* * *

The advent of the Cold War revived Canadian concerns for coastal defence. Beginning in 1948, Lancasters were retrieved from storage and issued to a succession of squadrons on both coasts. Lockheed P2V Neptunes began to supplant the Lancasters in March 1955, although the old bombers did not fade from Maritime Air Command until April 1959. Their service actually overlapped that of another four-engined patrol bomber, the Canadair Argus, which entered service in 1958.

Although MAC aircraft were primarily concerned with hunting submarines, they were involved in many other tasks, including ice patrols, search and rescue and Arctic sovereignty patrols. An aging MAC Lancaster, *Zenith,* was the first Canadian aircraft to fly over the geographic North Pole, in May 1949.

* * *

Canada's role in NATO air defence may best be illustrated by reference to correspondence that circulated within Air Force Headquarters late in 1950. Questions had arisen about the Soviet Union's bombing capabilities given estimates that by 1952 the USSR would have *1,000* TU-4 bombers (Soviet copies of the Boeing B-29) plus an array of light twin-engined jet bombers. While the TU-4s were capable of reaching targets in North America, the threat was limited because they could do so only by flying one-way (i.e., suicide) missions with a maximum 10,000-pound load. Britain, however, was vulnerable to raids by both the TU-4s and the many Soviet light jet bombers based in Eastern Europe. The Russians, it was felt, might fly 'night and bad weather missions employing pathfinder techniques and blind bombing methods' — although there was no evidence that they actually had the radar sets needed for such raids!

Working with these assumptions, the planners concluded that Canada needed an air-defence force of nine squadrons equipped with all-weather fighters and that Britain's defences needed bolstering with nine additional RCAF squadrons. However, by early 1951 the Canadian air commitment to NATO had been modified; the RCAF was to assist in the defence of Western Europe, deploying twelve squadrons.

This commitment began to take shape in October 1951 when personnel of 410 Squadron sailed from Halifax; their Sabre fighters were carried aboard HMCS *Magnificent.* On 1 November, RAF Station North Luffenham in England was officially transferred to 1 (Fighter) Wing, RCAF; No 410 and its aircraft took possession on 15 November. They were joined in February 1952 by 441 Squadron, which also travelled by sea, *Magnificent* once more serving as a transport. A different mode of delivery was used with the next unit. Between 30 May and 15 June 1952, 439 Squadron flew the Atlantic via Newfoundland, Greenland and Iceland. With their arrival, 1 (F) Wing was complete.

An advanced echelon of 1 Air Division Headquarters was established in Paris in August 1952. By year's end three more Sabre squadrons (416, 421 and 430) crossed the Atlantic. They constituted 2 (F) Wing at Grostenquin, France, and were the first RCAF units based in Europe since March 1946. No 1 Air Division Headquarters itself formed in Paris on 1 October as part of 4th Allied Tactical Air Force. The location was temporary; in April 1953 it was moved to Metz.

In the meantime more Sabre units were arriving in Europe. On 7 March 1953, Nos 413, 427 and 434 squadrons departed Canada; upon arrival at Zweibrucken, Germany, they constituted 3 (F) Wing. Nos 414, 422 and 444 squadrons made the trip between 27 August and 4 September; they became 4 (F) Wing, Baden-Soellingen. Finally, early in 1955, 1 (F) Wing moved from North Luffenham to Merville, France. The Air Division was now complete and it constituted the largest RCAF fighter force ever assembled.

The formation was never static, however. It began with Sabre 4 aircraft, powered by American J-47 engines. Early in 1954 the first Orenda-engined Sabre 5s arrived. Eventually improved Sabre 6 fighters were added. These machines had excellent performance, which contributed to high morale. Canadian Sabre pilots regularly 'bounced' and outflew their rivals in other air forces. On three occasions RCAF Sabre teams won the Guynemer Trophy, emblematic of gunnery supremacy in NATO air forces. The Sabre years were the happiest times for those serving in 1 Air Division; in retrospect, many would consider it the RCAF's golden age. In these years the RCAF was also at its peak post-war strength — 55,700 in 1958. However, even in peacetime there were casualties: 106 RCAF Sabre pilots were killed in accidents in Canada and Europe. Regrettably, one pilot defected to the Soviet *bloc* complete with a Sabre 6!

As NATO air forces grew (most notably after West Germany was admitted to the alliance), the role of 1 Air Division began to change. With day-fighter strength expanding, the RCAF was asked to reduce its Sabre strength and deploy all-weather fighters in their stead. There followed four mass transatlantic delivery flights in which four CF-100 squadrons were moved from Canada; the Sabre squadrons they replaced were relocated in Canada and re-equipped with CF-100s in their turn.

Replacement of the Sabre, commencing in 1961, coincided with a drastic change in Canada's NATO commitment. Selection of the Lockheed F-104 (produced in Canada as the CF-104) brought industrial benefits; Canadair of Montreal ultimately built 1,791 for the RCAF and for NATO allies. With the adoption of the high-performance CF-104 (described by some as 'a missile with a man in it') and disbandment of the CF-100 squadrons based in Europe, 1 Air Division was reduced from twelve squadrons to eight (later further reduced to six). However, when the Diefenbaker government accepted the CF-104 for NATO duties, it also accepted a role that was later deemed distasteful. Robert McIntyre has described the dilemma that beset not only 1 Air Division but the RCAF in general:

> Canada planned to use their CF-104s as nuclear strike bombers based in Europe for its NATO commitment. This role could only be effective if used in a first strike scenario — i.e. if NATO attacked the USSR in a surprise strike, a very unlikely happening. If the Soviets struck first, the 104 bases in Europe would be wiped out on any first strike, as they would be prime targets due to their nuclear role and close proximity to Soviet borders. The USSR would have little option but to use nuclear weapons in a war because of the Canadian role, hence negating any chance of a conventional war in Europe. One could argue that the CF-104 bomber force was a deterrent, as it narrowed the Soviets choice to either a nuclear war or no war at all. One can only conclude that the European CF-104 operation at that time was more political than practical in value...With the Bomarcs at North Bay and La Macaza, the CF-104 in Europe and the nuclear Genie missile equipped CF-101, Canada's aerial defence and strike capabilities were virtually all nuclear.[3]

The CF-104 was of course part of the nuclear debate of 1961–63 and the Air Division was issued nuclear weapons once the Pearson government had come to power. Nevertheless, even the Liberals were uneasy with the nuclear role and they set about quietly negotiating new tasks. By 1972 the Air Division had regained non-nuclear purity by taking on conventional strike and reconnaissance tasks; when the CF-18 Hornet began replacing the CF-104 in the mid-1980s it was non-nuclear from the outset.

While the Air Division was changing from Sabres to Starfighters (and from its conventional to a nuclear capability), the formation's commander, Air Vice-Marshal D.A.R. Bradshaw, gave evidence before a Parliamentary Committee in which he described the demands of the new role on personnel:

> Each pilot must attain and then maintain, a combat ready status to meet the high standards laid down by 4 ATAF and the RCAF. He must also satisfy the stringent nuclear and safety criteria dictated by the USAF. Finally, he must be able to

fly his aircraft at very low level, over devious routes, day or night and in all weather, up to distances of 200 to 600 miles and then deliver his weapon with pinpoint accuracy. He must then return safely to his base.

In contrast to our Sabre pilots, who flew in formation at great height with one or more other aircraft, assisted by ground radar for interception and navigation purposes, the strike pilot must carry out his mission alone, usually at low level and without any outside help. Therefore it becomes dramatically clear why the strike pilot must be so highly trained and have such a high degree of initiative and determination if he is to be successful in his task.[4]

Surprisingly — for those who think in terms of Hollywood images — the typical CF-104 pilot was thirty-three years of age, was married with two or three children and had accumulated 3,600 flying hours — 2,200 on jets, of which 225 were on the CF-104. (This of course at a time when the airplane was just entering RCAF service.)

Under NATO agreements, Canada undertook training of NATO aircrews in a scheme reminiscent of wartime BCATP operations. On 14 May 1951 the first group of NATO navigators received their wings at Station Summerside in Prince Edward Island. By 31 March 1952 the RCAF had trained 156 NATO pilots and eighty-four navigators and there were a further 1,166 NATO aircrew undergoing training. Most were British (852) or French (192), but there were also Dutch, Belgian, Italian, Norwegian, German and Danish aircrew. Early in 1953 Portuguese students joined in, while Turkish pupils arrived later in the year. By February 1957 some 4,500 pilots and navigators from ten NATO countries had been trained. The programme was then about to be reduced, as many of the countries involved had since built up their own training facilities. However, limited aircrew training continued for Denmark, Norway, the Netherlands and Germany.

* * *

A problem confronting Canadian planners soon after the Second World War ended was defence of the north — and how it was to be shared out between American and Canadian forces. From the very outset of the Cold War, USAF officers urged extensive continental warning and interception systems; their approach has been described as 'messianic and demanding.'[5] Air Marshal Leckie thought them alarmist and told the Cabinet Defence Committee as much on 13 November 1946. The Americans were placated for a time with limited co-operation, notably a 1947 joint defence policy that allowed their aircraft — including heavy bombers — uncontested passage through Canadian airspace. Canada established an Air Defence Group in December 1948 (elevated to Command status in June 1951) and RCAF fighters routinely flew exercises alongside American units.

The limited partnership worked only so long as the Soviet threat appeared remote. When the Russians exploded their first atomic bomb, pressure on Canada mounted for more tangible co-operation. In June 1950 the government agreed to the construction of a radar chain across the continent. This was the Pinetree Line, the first of three such networks to be built on Canadian territory — each more complex than the last and each becoming obsolete as the Soviet threat changed from piston-engined bombers to jet bombers to intercontinental ballistic missiles.

The original Canadian response to domestic air defence needs was to entrust the task to rejuvenated auxiliary units. Composed almost entirely of Second World War veterans, the auxiliary units initially resembled a club. They were also involved in exercises that closely resembled the fighter and tactical air-support operations of the recent war; for those joining the RCAF for excitement, the auxiliary of 1946–48 was probably more interesting than the regular Force. The reserve squadrons flew both older aircraft (Mustangs, Mitchells) and more modern types (Vampires until 1955, supplemented and then succeeded by Sabres).

As early as 1945 the RCAF had begun thinking of a long-range jet interceptor, preferably from a Canadian source and built to RCAF specifications. The desire for a domestic product

A munitions and armament technician loads rockets into a CF-100's wingtip pods, Weapons Practice Unit, Cold Lake. (CFP, PL 75386)

C-119s on tarmac, Abu Suweir, Egypt; Finnish troops disembark. (CFP, PL 106264)

C-130B Hercules ferry freight from Thule, Greenland, to Alert, Ellesmere Island, in the course of semi-annual resupply operations known as Operation BOXTOP. (CFP, PL 78070)

Yukon of No 437 Squadron in the Congo, 1960. (CFP, PL 138835)

was reinforced by the conviction that no British or American fighter, existing or planned, met Canadian needs. As of 1946–47 it appeared that any aerial threat to North America would come from the Soviet Union in the form of TU-4 bombers. The RCAF thus defined its needs, including the capacity to operate over a wide temperature range (-57 to +45 degrees C), 650-nautical-mile (1,196-km) radius of action, maximum speed of at least 490 knots (885 km/hr) at 40,000 feet (13,120 m). A new company, Avro (Canada), took up the challenge and thereafter developed the CF-100 — a conservative design that drew little upon wartime research at a time when German research into swept wings and compressibility was widely distributed among Allied aeronautical engineers. Tooling up for the manufacture of the first prototype began in January 1948.

As the Cold War intensified, RCAF planners found themselves studying documents that envisaged many potential scenarios, including a possible campaign to retake Iceland from invading Soviet forces. An 'Ultimate War Plan' drawn up in 1952 suggested that the RCAF might be built up to nearly 300,000 personnel, manning 116 squadrons at home and overseas, including Valiant, Canberra or B-47 bombers! The Canadian northern defence dilemma of protecting national interests was difficult, involving as it did a bilateral junior partnership with the United States. It was never thought that the continent could be turned into an impregnable fortress. Even deploying fighters equipped with guided missiles, the most optimistic expectations were that thirty per cent of attacking bombers could be destroyed. This would still be unacceptable in a nuclear war, so the ultimate defence was accepted as being a strategic bombing force that would deter such a war.

In 1953 the CF-100 began entering RCAF service. Ultimately it equipped thirteen squadrons, nine of which were deployed in Canada throughout the decade. Designed for Canadian conditions and built wholly in Canada, the 'Clunk' was also part of a Canadian industrial strategy. However, its intended successor, the CF-105 Arrow, proved to be far too expensive for the Canadian budget, especially in the absence of export markets that would reduce the unit cost. The Arrow was cancelled in February 1959, a decision hotly debated to this day.

Cost aside, one rationale for Arrow cancellation was that 'the day of the manned bomber is dead'. RCAF and USAF authorities argued strongly to the contrary and replacements for the slow, aging CF-100s were duly ordered — the CF-101 Voodoo (which began entering service in 1961) and the Boeing Bomarc-B surface-to-air missile. Both raised problems because of their nuclear punch. The Progressive Conservative administration of John Diefenbaker hesitated to take the final step and fit them with the atomic missiles and warheads that had been designed for them.

Meanwhile the RCAF had become ever more intimately involved with American authorities. Successive radar chains had been established across the country — the Pinetree Line (early 1950s), Mid-Canada Line (mid-1950s) and Distant Early Warning Line (DEW Line, constructed in the late 1950s). The DEW Line was particularly awkward, as some radar sites on Canadian territory were treated by USAF authorities as virtually American installations, to be visited only with USAF permission. This raised the most serious questions about Canadian sovereignty in the north since the construction of the Alaska Highway.

At the same time Canadian radar stations and home-defence fighter squadrons became ever more integrated into a North American system through the NORAD (North American Air Defence) Agreement signed in May 1958 and brought into effect on 1 August 1958. The Liberals had initiated negotiations for NORAD but had hesitated to implement them; Diefenbaker and his minister of national defence, George Pearkes, signed the accord without benefit of full briefings from either the chairman of the chiefs of staff committee (General Charles Foulkes) or the most senior officials in External Affairs. The NORAD agreement proved to be a poisoned chalice for Diefenbaker:

> By his unwitting consent, Diefenbaker had approved a tightly centralized defence system. At Colorado Springs the NORAD commander (or his Canadian deputy) could order Canadian and

American forces into action. Air warfare technology hardly allowed for prolonged reflection but only when he faced Liberal critics in Parliament did Diefenbaker realize the full implications. If politicians such as Robert Borden and Mackenzie King had struggled for Canada's right to control its destiny, Diefenbaker had unwittingly signed away his country's control of when it would declare war.[6]

The reckoning came in October 1962 during the Cuban missile crisis. American emissaries had expected that Canadian authorities would put Canadian forces on full alert; the Diefenbaker Cabinet debated two days before approving such action. Rightly or wrongly, the Canadian government had dared suggest that the senior ally might be mistaken or overly hasty. In fact, however, the Canadian forces (air and sea) were placed on alert (with approval of the minister of national defence but not with Cabinet consent) simultaneously with their American counterparts. Canadian commanders had chosen foreign direction over instructions from their own government.

The next step followed from the first; the question of nuclear arms for weapons systems already in Canadian hands had to be resolved. The Cabinet was split: defence minister Harkness favoured acceptance; external affairs minister Howard Green (a strong advocate of disarmament and arms control) was opposed. So long as the Liberal opposition kept silent (and by implication approved the government's inactivity) the issue might be swept under the rug, but American authorities would not allow that. In January 1963 General Lauris Norstad (retiring commander of NATO) declared at an Ottawa press conference that unless Canada accepted nuclear warheads for such weapons as the Honest John surface-to-surface missiles (sitting helplessly with the Canadian Infantry Brigade in Europe) she would be in default of alliance commitments. A political storm erupted: Diefenbaker denied the presence of nuclear commitments and Washington virtually called him a liar; the Liberals abruptly came out in support of 'honouring alliance commitments' (i.e., accepting nuclear weapons), which in turn split the Tory party and precipitated an election fought as much on the issue of nationalism versus continentalism as over conventional versus nuclear weapons. Continentalism and 'nukes' won; Lester Pearson's Liberal government honoured the earlier agreements, then set about quietly negotiating away the hated nuclear arms. This was largely achieved by 1972, the exception being the Genie missiles on the CF-101s.

* * *

Elsewhere the RCAF was working in new and old roles that served itself as much as they served the public. Chief among these was 'peacekeeping,' an offshoot of the United Nations Emergency Force created in November 1956 following the Suez conflict. Over the course of many such operations, three patterns of support emerged. The first was airlifting troops and supplies to troubled areas, first with C-119 Flying Boxcars and North Stars, later with Yukon transports, which in turn would be replaced by Boeing 707s. The second was establishment of a local Air Transport Unit operating lighter machines (Otter, Caribou and Buffalo STOL aircraft), which shuttled truce observers about places as diverse as New Guinea and Yemen, risking fire from nervous tribesmen, trigger-happy soldiers and suspicious fighters. On 17 May 1967, for example, an RCAF Caribou was harassed and fired upon by Israeli aircraft over the Sinai desert. The third pattern of support was RCAF management of air transport even when no Canadian aircraft were participating. This was most evident in the Congo, 1960–61, when Group Captain W.K. Carr, with a staff of ten, directed air transport operations over a land mass equal to one quarter of Canada. They also dealt with problems inherent in coping with thirteen types of aircraft, eight languages and local political instability.

Nothing gave the RCAF a better image than aerobatic teams, as the former Siskin flight had shown. In 1949, 410 Squadron formed one, the Blue Devils. Its pilots were described as 'solid wartime fighter jocks'. Through 1949–50 the team, flying Vampires, performed at such events as the Michigan Air Fair, the Cleveland Air Races and the Canadian National Exhibition. They disbanded when the squadron converted to Sabres and proceeded overseas. Another group, the Sky Lancers, was formed within 1 Air Division and performed at European air shows in 1954 and 1955. Rehearsals for a third season ended in a terrible crash on 2 March 1956; the four-plane team executed a loop, failed to pull out in time and thundered into the ground and exploded, in formation to the end.

To celebrate the fiftieth anniversary of flight in Canada, the RCAF formed another aerobatic team — again with Sabres — dubbed the Golden Hawks. They were authorized to operate for one year only but proved so popular that they remained on the air-show circuit until the end of 1963. During that time a group of Harvard instructors formed a team — the Goldilocks — that mocked the fancier Sabre team. The Golden Hawks were disbanded as an economy measure. For Centennial year, 1967, the RCAF formed a new aerobatic team, the Golden Centennaires, this time with CL-41 Tutor trainers. That group, too, proved more popular than expected; disbanded the following year, they were quickly revived as the Snowbirds and continue to perform across Canada with fresh pilots and aging jet trainers.

* * *

Changes were to be expected; not all were welcome. On 1 August 1964 Air Marshal C.R. Slemon, the last of the original 1923 class of provisional pilot officers, retired. His departure coincided with policy shifts that were at times debatable if not peculiar. One such instance, in 1965, was the release of 500 experienced pilots and air observers, announced as an economy measure. Within a year the force was short of trained aircrew and attempting to re-enlist some of those who had just been let go.

Beginning in 1964, the Liberal government undertook a drastic re-organization of the Canadian forces. The first step was the creation of a single command structure at National Defence Headquarters; the positions of chief of the general staff, chief of naval staff and chief of air staff were abolished. Air Marshal F.R. Miller became the first chief of defence staff presiding over a unified command. The next step affected operations directly; eleven field commands across Canada were merged into six — Mobile, Maritime, Transport, Air Defence, Material and Training. Next, in 1966, came the functional unification of the armed forces, with single-purpose recruiting centres, single-purpose specialist training centres (as in electronics, food preparation, etc.) and a unified, automated pay structure. A new National Defence Act, passed in April 1967, led to the final unification of the forces, effective 1 February 1968. These latter steps coincided with the introduction of new service dress uniforms common to sea, land and air elements — uniforms that became best known as 'Jolly Green Jumpers' (one of the more printable tags).

Not surprisingly, integration and unification were not universally welcomed by the forces. Opposition was most vocal and public within the navy. The air force, lacking such things as regimental traditions comparable to those in the army, undoubtedly had the fewest doubters. Nevertheless, there were reservations at senior levels. However, those with misgivings retired quietly rather than resort to public statements. The RCAF had ceased to exist, but there was still an air force.

Chapter VIII

THE NEW AIR FORCES

With unification, the RCAF ceased to exist as a formal organization. The air force that did emerge, however, constituted a successor to the three former air arms. A review of earlier events will explain the process.

The post-war Royal Canadian Navy had sought to form a 'balanced fleet' rather than to become a scaled-down version of its wartime self. Aircrew training under Royal Navy and United States Navy auspices had begun in 1942. The RCN then set out to acquire aircraft carriers; two was the preference, but the post-war establishment of 10,000 men ruled out such a plan and the RCN would henceforth make do with one carrier at a time — *Warrior* (1946–48), *Magnificent* (1948–57) and *Bonaventure* (1957–76).

In the late 1940s and early 1950s the RCN scarcely resembled what it had been during the Second World War and what it would become under NATO — a specialized anti-submarine force. The vessels that sailed as escorts to the carriers and the mock battles they fought would have been appropriate to the operations anticipated off Japan had the Japanese not capitulated in August 1945. In March 1950, for example, *Magnificent*'s aircraft manoeuvred off Cuba with American and British ships, conducting air strikes so vigorously as to remind some American officers of carrier actions in the Pacific war.

The RCN air arm used a succession of veteran British aircraft designs — Seafires, Fireflies, Sea Furies — but ultimately came to rely on more current American types. From 1952 to 1957 Grumman Avengers served as carrier-borne submarine hunters; they were replaced by Grumman Trackers built under licence in Toronto. McDonnell F2H Banshee jet fighters entered service in 1955. Unfortunately the Banshees were too fast and heavy to be accommodated on *Magnificent*, and aboard *Bonaventure* they could be operated only under the most favourable weather conditions. '*Bonnie*' normally handled eight Banshees and twelve Trackers, but as early as May 1958 senior officers complained that neither figure constituted a viable force, either as an anti-submarine unit or as a defensive fighter flight. The Banshees were more significant in their assignment to NORAD defences where their fast rate of climb and Sidewinder missiles made them formidable interceptors.[1] The Banshees were retired in 1962, even before '*Bonnie*' would be paid off in 1970. That move reduced the naval air component at sea to helicopters only. Fixed-wing naval aircraft such as the Trackers became shore-based sentinels employed in fishery patrols and occasional search and rescue (SAR) missions.

The Navy brought its own legacy of aviation history to the unified forces, including intrepid mercy flights. A Sikorsky H-55 helicopter dubbed the 'Shearwater Angel,' had a particularly interesting career between June 1955 (when it was acquired) and May 1970 (when it was transferred to the National Aviation Museum). It served aboard both *Magnificent* and *Bonaventure* and operated as far away as the Sinai peninsula. It also participated in several medical sorties, delivering supplies or evacuating patients from isolated communities, and in eight rescue missions at sea; the latter included saving twenty-one sailors from a sinking freighter on 26 November 1955.

* * *

Naval interest in air operations was matched by similar concerns in Army circles, which sought either to have aircraft under direct Army control or to secure closer co-operation with the RCAF. Army air-mindedness increased following discussions in 1946, when the RCAF was contemplating acquisition of new transport aircraft; Army officers were concerned that their requirements might not be factored into Air Force specifications. Early in 1947 senior Army officers suggested that a committee be established in National Defence Headquarters to co-ordinate the air policies and operations of all three services. The air staff agreed to a team composed of middle-ranking officers; the Army pressed for (and got) a more powerful body.

In April 1947 the Joint Air School (later named the Canadian Joint Air Training Centre) was formed at Rivers, Manitoba, so that the Army and Air Force could work more closely together in any future conflict. While the first task would be instructing ground forces in air portability (including parachute delivery), staff training would also take place; regimental officers and senior NCOs were to learn not only the techniques of land/air warfare but also how to teach those skills to others, including Reserve personnel.

A manifestation of the Army's ambitious thinking was the formation of a Special Air Service Company — a predecessor of the Canadian Airborne Regiment — but the Company had only a short life before its eventual absorption into the Joint Air School. As a result, apart from isolated air observation post (AOP), helicopter training and communications flights, Army aviation continued to be overshadowed by the RCAF.

The Army's training and exercises were governed to a great degree by its most recent wartime experiences. In 1946–47, thirty-two Waco CG-4 gliders and two other Waco variants were distributed to Active Force units; most had been erected at Rivers. They were used periodically in trials and training. For example, in February 1951 men of the Royal Canadian Regiment and the Royal Canadian Artillery were dropped by parachute and glider to repel an 'enemy force' in the form of the Royal 22e Régiment entrenched eighty-five miles north of Churchill, Manitoba. Nevertheless, gliders represented obsolete technology; helicopters and heavy transport aircraft like the Fairchild C-119, capable of air-dropping jeeps and light artillery, would render them redundant, and the last gliders were scrapped in 1955.

Another move by air-minded soldiers was the development of an independent AOP capability similar to that utilized in April and May 1945. The peacetime Army proposed to continue with its own artillery co-operation units, while training of Army pilots was a function of the Joint Air School from the outset. Thirty-six Auster VI aircraft were acquired from Britain and were used by both wartime AOP veterans and a new generation of Canadian Army pilots.

While the Army used light aircraft chiefly in the AOP role, the machines had other uses as well. Thus in October 1948 Austers were used at Camp Wainwright, Alberta, to place an umpire

Auster VI representing the army's new interest in aviation. (CFP, RE 64-1844)

Avenger over Chebucto Head, Halifax. (NAC, DNS 3288)

Two T-33 Silver Star trainers over Halifax. Although closely identified with the RCAF, these are RCN machines. (NAC, DNS 14721)

over armoured and infantry units during exercises. They also enabled a safety officer to supervise the manoeuvres while live ammunition was being fired. The Austers provided tactical reconnaissance for all parties involved. On two occasions they became an enemy air force, carrying out mock strafing and bombing runs to familiarize recruits with these tactics. The bombs were one-pound flour bags identical to those used in the 1920s and 1930s during RCAF-Army co-operation training.

Army fliers were committed to battle during the Korean War. Several officers served in 1903 Independent Air Observation Post Flight, which acted as the eyes of the Commonwealth Division, and Captain Peter Tees, RCA, was awarded a Distinguished Flying Cross for having flown 185 sorties and directing 453 artillery 'shoots.'

Army air work, whether in Canada or Europe, had its own peculiarities. The emphasis was on low-level flying, using ground features to avoid detection. Most of the related training was around Rivers, an area so flat that finding decent obstacles to hide behind was a challenge in itself. Undaunted by unco-operative terrain, Army pilots skimmed as low as six feet (two metres), using ravines and trees to screen their movements. In NATO exercises, these hill-and-gully tactics were co-ordinated with tank and scout-car movements.[2]

Long before formal unification, the various air elements had shared many training facilities, including aircrew selection offices at Centralia, Ontario, and fixed-wing primary training at Portage la Prairie, Manitoba. A few aircraft types were common to two services at any one time; Expeditors and T-33s could as easily be RCN as RCAF machines. CH-113 helicopters were flown by both the Army and the Air Force and joint operational training in cargo helicopter flying was conducted at Trenton, Ontario.

The creation of No 10 Tactical Air Group in September 1968 demonstrated how closely the formerly separate air forces were now being redefined. It included two CF-5 fighter squadrons plus T-33 reconnaissance aircraft, Buffalo tactical transports and a variety of helicopters for transport and observation duties across the country. Significantly, 10 TAG was viewed not as an Air Force formation but as a component of Mobile Command. Eventually it would become entirely a helicopter formation, its fixed-wing aircraft being transferred elsewhere.

The forces retreated slightly from unification in September 1975 upon the establishment of Air Command, with Lieutenant-General William Carr responsible for all military air forces. This included 10 TAG, although the latter remained firmly tied to Mobile Command's planning and exercises. The annual departmental report described Air Command as 'providing greater flexibility in the employment of air power,' and added, 'In the same way as sailors and soldiers relate to Maritime and Mobile Commands, Air Command now is the focal point of tradition and professional expertise for airmen of the Canadian forces.' Although the RCAF itself was not revived, the new formation took as its motto the *'Sic Itur ad Astra'* formerly used by the Canadian Air Force of 1920–24. In February 1985 the minister of national defence announced reintroduction of distinctive naval and air force uniforms. The first 'air force blues' were unveiled that year and by 1988 the CAF air element looked very much like the pre-unification force.

* * *

It is easy to forget the Cold War, but its presence and its changing nature were constant factors until 1989–90. The Cuban missile crisis of October 1962 marked a low point in East-West relations. It was followed by successive thaws and chills. Treaties to restrict and then to ban atomic tests, as well as superpower Strategic Arms Limitation Talks (SALT), promised relaxation of tensions; Soviet intervention in Afghanistan (1979) and the destruction of Korean Air Lines Flight 007 by Soviet fighters (1983) raised fears and provoked angry rhetoric on both sides.

Canadian military spending remained high during the Cold War, although the increasing costs of equipment meant that personnel resources were stretched. Thus although the Department of National Defence budget rose from $1,824 million in 1970–71 to $7,840 million

An RCN F2H Banshee on armament exercises at the Joint Air Training Centre, Rivers, Manitoba. (CFP, DNS 26433)

CF-104 on low-level bombing practice, Sardinia. (CFP, PMR 98-170)

in 1983–84 and then $12,005 million in 1990–91, the number of Regular Force personnel remained surprisingly stable — 90,100 in 1971, falling to 78,150 in 1975, increasing to 82,000 in 1983 and to 89,000 in 1991. The air forces held their own in numbers; when Air Command was created in 1975 it had 22,800 service personnel (Regular and Reserve) plus 7,800 civilians. The 1990 figures were 20,600 and 5,700, respectively. If the forces seemed smaller, it was due in large measure to highly publicized base reductions, such as the closure of Canadian Forces Base Summerside (Prince Edward Island), and shrinkage of units to a point where maritime patrol squadrons that had earlier boasted a complement of a dozen Argus aircraft now operated with five or six Aurora machines; expansive tarmacs looked increasingly empty.

The post-integration forces were subjected to near-constant scrutiny as policy white papers succeeded one another, interspersed with ministerial reviews and parliamentary inquiries. A 1971 white paper paved the way for cutting NATO commitments while converting CAF air units to a conventional (as opposed to nuclear) strike role. The 1987 white paper carried a particularly grand title, *Challenge and Commitment: A Defence Policy for Canada,* and promised expensive re-equipment of the forces. Nevertheless within two years defence budgets had been reviewed and many of the intended purchases cancelled.

In practice, each examination brought either modest reductions or re-organization, sometimes with peculiar results. Thus soon after initial deliveries of CF-5 tactical fighters had been made to Mobile Command squadrons, many of the new aircraft went directly into storage. From 1968 through 1985 the CF-5s served in various roles, based in Canada but ready to move overseas in fulfilment of changing NATO needs. In the mid-1980s they became training aircraft for aspiring CF-18 Hornet pilots. By 1995 the forces were training only fifteen to eighteen CF-18 pilots per year and the CF-5 fleet was retired to save $100 million annually. Similar cuts to programmes and aircraft were as bewildering as they were debilitating.

Old RCAF hands were pained as new phrases entered the service vocabulary, phrases like 'corporate policy framework' and 'Alternate Service Delivery' that were euphemisms for planning and privatization. An example of the latter was the sale of CFB Portage la Prairie to a civilian corporation, Southport Aerospace Centre, Inc. (SACI), which turned the site into an aero-industrial park. Meanwhile the Canadair corporation was awarded a CF Contracted Flying Training and Support agreement and thereafter used SACI facilities to deliver primary, helicopter and multi-engine training for CAF personnel. Still, these developments were not unique to Canada or the Canadian Forces: Australia, Britain and New Zealand (among other countries) adopted similar approaches to non-operational activities.

The armed forces (air elements included) have continued to be active in Aid to the Civil Power in unprecedented ways. Operations chiefly involve troops, with aircraft serving primarily to get soldiers and their equipment to the area affected, but aviation has contributed something unique in almost every instance. Thus during the 1970 October crisis CF-5 fighters flew photo-reconnaissance missions south of Montreal, searching for unusual activity that might indicate a Front de Libération du Québec hideout. Mindful of the terrorist attack that had killed Israeli athletes at the 1972 Munich Olympic Games, the Canadian government deployed substantial forces to assist with security for the 1976 Games in Montreal. Again, troops were the most visible element, but they were supported by the constant presence of helicopters shuttling men about and conducting surveillance operations. Special air restrictions were imposed over all the Olympic sites. All aircraft entering these zones had to identify themselves and receive clearance from the Olympic security headquarters, and CF-5s were on standby to intercept any unauthorized aerial intruders.

At various times, prison riots or police strikes have brought similar CAF transport and surveillance work for periods lasting from one to four days. A sustained instance of Aid to the Civil Power was Operation SALON, better known as the 1990 Oka crisis. Indeed federal intervention in that affair began with an airlift that brought 150 RCMP officers from western Canada to aug-

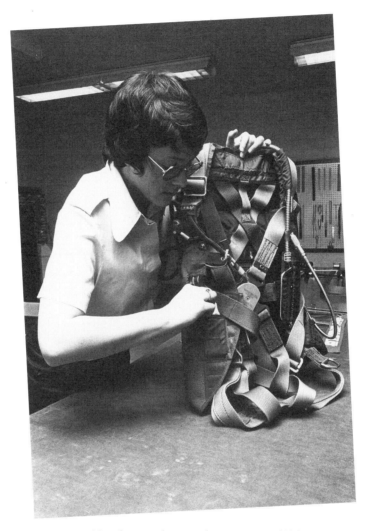

SS tech at work, 1977. (CFP, DHH 90/447)

ment the Quebec constabulary, and CAF helicopter crews constantly monitored the situation — including what was happening behind Native barricades.

More orthodox aid has been rendered in times of natural disaster. An example of such work was that done at St. Jean Vianney, Quebec, which was beset by landslides and cave-ins on 4/5 May 1971. The first CAF helicopter pilot over the site described a hole resembling a moon crater. Helicopters from CFB Bagotville and Petawawa participated in SAR operations while CF-5 and CF-100 aircraft flew photographic and reconnaissance missions, searching for other geological faults in the area. The region benefited from CAF air operations again in July 1996 when two Hercules transports and twenty helicopters evacuated more than 1,500 residents of Grande-Baie following disastrous floods in the Saguenay area.

These operations were eclipsed in scale by two massive assignments in 1997 and 1998. The first, appropriately named Operation NOAH, was assistance in dealing with the Red River flood. Air support took many forms, including airlift of troops and special equipment to Manitoba and helicopter sorties to supply isolated communities, sometimes evacuating people in critical circumstances. When a large new dyke had to be built quickly, CAF aircraft dropped parachute flares to provide night illumination for those constructing it. Two CP-140 Aurora maritime patrol aircraft with their sophisticated electronic systems served as communications relay links across the temporary inland sea. Assistance to civil authorities in Quebec and Ontario during and after a devastating ice storm in January 1998 did not require such unusual moves, but it nevertheless saw a major airlift, including movement of generators from across Canada to the stricken areas, and helicopter flights to reconnoître damaged power lines.

Search and rescue has been one of the most publicly recognized areas of forces work. Most sorties have been without incident, but on 2 November 1971 a CAF Dakota crashed while parachuting supplies to the crew of a downed civilian aircraft that had been provisioning the Cape Parry DEW Line station. Nine servicemen died in the accident.

In 1976 the Mynarski Trophy was instituted to honour yearly the most deserving Canadian organization or individual engaged in aerial SAR. It has become the most coveted honour among CAF units so engaged. The importance of SAR was further recognized in September 1979 when a new service trade was established, that of the SAR Technician. Most such operations have involved one or two victims, but there have been exceptions, such as the rescue of forty-two of the 128 passengers and crew from the Labrador ferry MV *William Carson* on 2 June 1977 after the vessel sank (the other eighty-six people were picked up by a ship). CAF helicopters plucked twenty-six crewmen from a grounded and storm-ravaged freighter near Sable Island, Nova Scotia, on 26 November 1981, one Sea King involved in the rescue being packed with thirteen sailors and a crew of four when it finally left the wreck. Two helicopters rescued twenty-seven crewmen from a foundering bulk carrier, the MV *Dahlia'D,* during a winter storm on 16 March 1982. One of the most exciting stories in the recent history of the Canadian Forces was the rescue effort conducted near Alert, Northwest Territories, in October 1991 following the crash of a Hercules transport, an epic that has since been the subject of a book and a film.[3]

A year-by-year accounting of SAR activities would be repetitious, but one year's statistics may be taken as typical. In 1985 the four Rescue Coordination Centres, with headquarters at Halifax, Trenton, Edmonton and Victoria, reported 8,309 incidents. Of these, 6,334 involved ships in distress, 1,236 related to aircraft and 553 were humanitarian episodes requiring aerial evacuations; 186 incidents involved assistance to civilian authorities.

The forces have long been prepared to deal with emergencies in high latitudes, honing skills in the course of less perilous though no less significant tasks. Between 22 April and 2 May 1974 Air Transport Command joined forces with a scientific research team in a most unusual exercise. Two Labrador helicopters escorted by two Hercules transports carried a diving party of scientists to the North Pole. The immediate task

Sikorsky helicopters, No 50 Squadron, Knob Lake, Quebec, marshalling to supply eastern Mid-Canada Line radar sites. (CFP, RCN 0-8999)

CH-113 on rescue operation. (CFP, PMR 98-166)

was to conduct experiments under the ice, but another objective was to test the capability of the forces to deploy SAR units in the high Arctic in response to an air disaster.

In March 1976 the first of several northern Major Air Disaster (MAD) exercises was held, with thirty-five army cadets acting as crash victims located by a Hercules and 'evacuated' by ski-equipped Twin Otters.[4] A similar operation in March 1994, dubbed SAREX 94, was the first such exercise to involve American, Canadian *and Russian* specialists, dealing jointly with a simulated airline disaster in Alaska. Given the international makeup of the exercise, there were plenty of resources on hand. One participant reported, 'for a while, our landing zone was busier than Pearson International Airport as choppers of all sizes and description landed, loaded and departed for the hospital.'

Although not a mercy mission, Operation MORNING LIGHT (24 January to 22 March 1978) tested not only search skills but also the function of supplying ground forces in the field. MORNING LIGHT was a response to the breakup of a Soviet space satellite, COSMOS 954, over the Northwest Territories. The spacecraft had been powered by a nuclear reactor. Although subsequent operations were described as being aimed at recovery and disposal of dangerous atomic waste, the possible intelligence value of satellite fragments was a Cold War factor. Specially equipped CAF aircraft scoured the satellite's last known track from Great Slave Lake northeast towards Baker Lake. The first radioactive ' hit ' came on 26 January; actual debris was found two days later. Each discovery was followed up by recovery teams, who removed fragments and cleaned up each site as best they could before spring melts could disperse the material. MORNING LIGHT was carried out amid winter storms common in the area.[5]

Mercy flights to distant countries have also been part of CAF aerial operations. Two factors have increased the scale and number of these tasks — the growing range and payload of heavy transport aircraft and the development of Short Take-Off and Landing (STOL) aircraft capable of lifting bulky cargoes. Scarcely a year has passed that a CAF Boeing 707, Yukon or Hercules aircraft has not been filmed delivering food or medicine to countries ravaged by famine, flood or earthquakes. The STOL crews, with less publicity, have operated in some of the world's most forbidding topography to distribute supplies to outlying communities.

A Peruvian earthquake in June 1970 demonstrated what both 'heavies' and STOL machines could accomplish. Yukons and Hercules airlifted 123 tons of equipment to Lima, including X-ray machines and diesel generators. Five Caribou aircraft flew to Lima, then established bases in the heart of the Andes, where the high elevations affected aircraft performance and short runways taxed their STOL capabilities. Aircrews accepted what they deemed to be prudent loads, then brushed treetops on take-off. They flew into canyons, fighting turbulence while dropping bundles to isolated villages. With skill and luck, the CAF crews operated with no casualties; four helicopters and one transport from other countries crashed in the Andes downdraughts.

CAF aircrews have so frequently deployed overseas on relief missions that Canadians tend to take them for granted, viewing them as standard players in nightly telecasts from troubled corners of the globe. Operation ASSURANCE, for example, is chiefly remembered as a short-lived attempt, in November and December 1996, to establish a multinational force in the eastern Congo — an operation that was abruptly halted when most of the refugees it was intended to help about-turned and returned to their homes in Rwanda. Nevertheless, Canada had been prepared to conduct a massive humanitarian operation. Of thirty CAF Hercules transports, twenty-five had been assigned to ASSURANCE, and Air Transport Control Units had been manned at Entebbe (Uganda), Nairobi (Kenya), Lyneham (England), Heraklion (Greece) and Djibouti. The effects of the deployment were felt in Canada, where Buffalo aircraft normally based in Comox, British Columbia, were reassigned to Manitoba, assuming SAR responsibilities normally assigned to Winnipeg-based Hercules transports.

Mercy missions, as well as military supply operations, have employed more than new aircraft; they have also involved some innovative techniques. None has been more spectacular

Caribou and Otter support UNEF. (CFP, PL 150543)

An Argus of No 405 Squadron, Greenwood, with a selection of No 54 depth charges. (CFP, PL 114511)

than the Low Altitude Parachute Extraction System (LAPES), by which an aircraft (usually a Hercules) skims barely more than a metre above a flat piece of ground while its cargo, strapped to a reinforced pallet, is yanked out the rear cargo door by parachute. The first such CAF operation, carried out on 17 July 1973, delivered fuel drums to an Arctic survey site. Unhappily, however, the system is as potentially dangerous as it is effective: on 16 November 1983 a ramp failure on a Hercules prevented a 30,000-pound (13,600-kg) load from clearing the aircraft, causing it to virtually stand on its tail until it stalled and crashed. Seven crewmen died.

* * *

NORAD defences remained a priority both before and after 1968, although the principal threat clearly shifted from manned bombers to intercontinental ballistic missiles (ICBMs). This was tacitly admitted in 1972 when the two Bomarc missile squadrons were disbanded. The Bomarc had been at the centre of the 1962–63 controversy that brought down the Diefenbaker government, yet it had always lacked the glamour of conventional fighters. All live firing training had been conducted in the United States, where the Canadians achieved excellent results, even scoring a direct hit on a radio-controlled Starfighter target. For ten years the Bomarc squadrons at North Bay, Ontario, and La Macaza, Quebec, had done nothing more menacing than occasionally raise their missiles to firing position before lowering them again into storage bays.

Fortunately Canada's post-1945 air defences have never been tested in war, although over the years successive probes by Soviet long-range bombers resulted in numerous scrambles by CF-100s, CF-101s and CF-18s. These games were mutually rewarding: Canadian aircrews practised their drills and training while Russian aircrews tested and assessed NORAD responses. The defenders were on alerts ranging from five minutes to an hour. One writer has described a typical scramble from the early 1980s involving Voodoos of 425 Squadron, Bagotville:

Once the alarm sounds, two Voodoos have to be airborne and en route to Gander within one hour. There they refuel and stand by to go out and fly Combat Air Patrol (CAP), which is sitting on station waiting for the Soviet 'Bear' bombers that occasionally test our defences. The 101s always travel in pairs for such intercepts, at 200 miles off Canadian shores, with no witnesses around, is not an ideal situation for a single aircraft to be playing war games with the Russians. It has been known to be fatal.[6]

Over the years NORAD forces changed and lessened. At its peak (circa 1963) NORAD had seventy-eight radar stations in the DEW Line, ninety-eight stations in the Mid-Canada Line and thirty-nine stations in the Pinetree Line. These were supplemented by various American sensor systems aimed at preventing an end-run around the main radar lines — Aircraft Warning and Control System (AWACS) aircraft, US Navy picket ships and offshore Texas Towers. These passive defences were backed by enormous aerial forces — 3,000 interceptor fighters (200 of them Canadian) plus ninety Nike and Bomarc missile squadrons (two of them Canadian). These were directed from computerized regional control centres known as SAGE sites (Semi-Automatic Ground Environment) dug into the Rocky Mountains (Colorado Springs) and the Canadian Shield (North Bay).[7]

By 1985, following three NORAD agreement renewals, the DEW Line had been reduced to thirty-one sites and the Pinetree Line to twenty-four, and the Mid-Canada Line had been closed since 1965. Most of the picket ships and Texas Towers were gone and anti-aircraft missile squadrons were down to nine (all Nike batteries controlled by the US Army). The total number of interceptors had declined to 300, of which thirty-six CF-101s were in CAF units. The emphasis had changed to warning of ICBM strikes using Ballistic Missile Early Warning Systems (BMEWS), the Satellite Early Warning System (SEWS) and the Space Detection and Tracking System (SPADETS). None of these new systems involved facilities based in Canada; only two Canadian sites were involved in the SPACETRACK system (radar and camera sites surveying the skies) and interceptors were now detailed to peacetime surveillance and interception of unidentified aircraft and limited bombers. This reliance upon passive defences made it easier for the Canadian government to escape its earlier nuclear role. Although the new fighter adopted in 1982 (the CF-18) was ostensibly intended for both NORAD and NATO duties, its acquisition was more a response to Allied NATO pressure than domestic air-defence needs.[8]

The continued shut-down of radar sites underlines how little the manned bomber was deemed a threat. In 1987–88 the Pinetree and DEW lines were closed. The bases still intrude upon the news as governments and environmentalists argue as to whether the sites have been adequately cleared of toxic wastes.

Post-integration forces lived with pre-integration commitments, including 1 Air Division's CF-104 Starfighters dedicated to a reconnaissance and nuclear-strike role. As of 1968 the formation had six squadrons, four of which were equipped to carry atomic weapons — although control of the trigger remained in American hands. Nevertheless air crews had to accept the fact that if they were ever scrambled to fulfil their mission their survival would be problematical; even if they did return, their base was likely to be a radioactive wasteland. One pilot summed up their peculiar situation: *'You had to realize, respect and yet not let it bother you, that that small cigar on the centreline position made Little Boy and Fat Man look like toys'*[9]*

As of 1970, 1 Air Division had been reduced to three squadrons and redesignated 1 Canadian Air Group. By 1972 it had assumed a conventional ground-attack role, on the assumption that a war in central Europe might somehow be kept non-nuclear for some time. The strategic assumptions associated with the role were questionable, and one pilot recalled, *'Some thought loading a thoroughbred like the '104 with iron bombs was like delivering pizza in a Rolls-Royce — not economical but certainly classy.'*[10] Whether in a nuclear or conventional role, CF-104 aircrews accepted great risks in low-level, high-speed exercises, both in Canada and overseas. A bird ingested into the engine could be fatal; even successful bail-outs from crippled aircraft were hazardous. Approximately 750 Canadian pilots flew the CF-104; thirty-seven were killed, including seven who died following ejections, usually when parachutes failed to deploy, although eighty-four CF-104 pilots bailed out successfully after incidents that ranged from engine failure through bird strikes to mid-air collisions.

There were times in the near past when our forces seemed to be serving everywhere at once. While Canadian troops and aircraft had concentrated upon defending the central European front in the first two decades after the formation of NATO, in the early 1970s additional commitments were made to assist Norway in the event of war, using Canadian-based units that otherwise would have reinforced the regiments and squadrons in Germany. The force committed to this northern endeavour was designated the Canadian Air/Sea Transportable (CAST) Brigade Group, which exercised regularly in Canada with brief biannual transfers to Scandinavia. Thus in September 1972 CAF Hercules transports moved a battalion from Canada to Norway, then brought them back, an operation requiring seventy-five transatlantic round trips. Having acquired in-flight refuelling capabilities with Boeing 707s, the CAST force undertook regular movements of CF-5s to Norway. These exercises continued well past the demise of the Cold War; the last, in February 1995, saw deployment of twelve CF-18 Hornets from Bagotville to Norway.

The Canadian government announced in 1975 that it was looking for a new fighter to replace the Voodoo, Starfighter and CF-5. A prolonged search followed, with several types contending, and the final choice was the McDonnell-Douglas F-18 Hornet. The key to its selection lay in its particularly rugged construction, stemming from its original design as a carrier-borne aircraft. Unlike many previous RCAF aircraft, the F-18 would have no Canadian production component; the machines would be built in the United States, although with certain modifications for Canada. These included an Arctic survival kit and a 60,000-candlepower spotlight for air-to-air identification at night. These changes were sufficient for a distinct designation — the CF-18.

---

* Little Boy and Fat Man were the atomic bombs dropped respectively on Hiroshima and Nagasaki.

The first delivered was to Cold Lake, Alberta, in October 1982. Conversion and training of units commenced soon afterwards, but only in May 1984 did the first CF-18s become operational with Canadian NATO squadrons. A total of 138 were acquired (ninety-eight single-seat, forty two-seat).

No 1 Air Division had comprised twelve squadrons in its Sabre days and eight squadrons when it converted to CF-104s. That number decreased over time — to six in 1967, then to three in 1970 when the nuclear strike role was eliminated. With CF-18 deployment, the number of Canadian NATO fighter squadrons remained at three, while in Canada four other squadrons operated the new fighter. Home-based Hornets occasionally joined their European counterparts to practise transatlantic transfers. Between August 1991 and November 1992 the three CF-18 squadrons overseas were disbanded as the government reduced NATO commitments; their aircraft were returned to Canada. As of 1997 two squadrons were permanently based at Bagotville and three at Cold Lake. Additional machines were on strength with the Aerospace Experimental and Training Establishment. From this pool, aircraft have been dispatched to wherever the Canadian government has wanted to show the flag.

Ever since the creation of NATO, Canadian facilities have been used to train Allied aircrews, either at Canadian bases or by providing facilities for combat exercises such as those currently staged over Labrador. Providing national airspace for air forces of densely populated countries whose aircrews would otherwise have no place to hone battle skills has been attended by protests and controversy. On the other hand, merging Allied personnel into CAF training programmes has never been a problem. When both Canada and the Netherlands adopted the CF-5 as a tactical fighter, it made sense to train personnel for both countries at Cold Lake. Between 1971 and 1983 a total of 282 Dutch pilots received their wings following Canadian-based training.

* * *

Canadian interest in peacekeeping operations as an integral element of foreign policy has ensured that forces would be widely scattered across the globe for periods varying from a few months, as in Zimbabwe in 1980 and Namibia in 1989, to several years, as in Cyprus. Most personnel have been troops engaged in supervising ceasefires, but air transport units have delivered them in Yukons, Boeing 707s and Hercules aircraft, while communications helicopters and light aircraft have shuttled personnel about within the theatre. In Haiti, from 1995 through to 1997, the primary CAF contribution comprised eight CH-135 Twin Huey helicopters (later replaced by CH-146 Griffons) assisting both the local government and UN forces scattered about the nation. It is worth noting that commitments such as these affect CAF operations at home. Thus in June 1996 provincial forestry officials requested twenty military helicopters to help combat fires in northern Quebec; 10 TAG was able to respond with only half that number due to the formation's overseas helicopter assignments, notably in Haiti.

Despite the peaceful objectives of UN operations, flying in support of them has had its hazards. In September 1965 an RCAF Caribou was destroyed on the ground by Pakistani aircraft during a clash between India and Pakistan, while in December 1971 a CAF Twin Otter was destroyed by Indian Air Force machines in similar circumstances. Neither incident involved casualties. On 9 August 1974, however, a CAF Buffalo was shot down by a Syrian ground-to-air missile, killing all nine persons aboard.

The end of the Cold War coincided with fresh troubles in the Balkans, and forces were marshalled by the UN, and later NATO, to intervene. Their aims were complex and confused, and Canadian participation reflected Allied uncertainties. At various times Canadian aircrews were operating helicopters, transports, maritime patrol aircraft and fighters in a war-torn region euphemistically described as the 'Former Yugoslavia.'

Balkan operations have been particularly frustrating and dangerous because the diverse belligerent forces have been unpredictable and undisciplined. When fighting closed Sarajevo airport to civilian air traffic in July 1992, Operation AIRBRIDGE began as a United Nations effort relying on aircraft from eleven

Bomarc missile. (CFP, PL 137377)

countries to succour the city. It was expected to last two weeks; instead, AIRBRIDGE became the longest-running UN airlift in history. On 14 July 1994 a CAF Hercules made the 10,000th delivery. AIRBRIDGE was stressful work; an Italian transport was shot down and UN personnel reported more than a hundred aircraft 'incidents,' chiefly involving search radars of one faction or another locking onto transports, a palpable warning of possible missile attack. Crews flying to Sarajevo approached at about 20,000 feet (6,450 metres) to remain out of the range of ground weapons. Close to their objective, they spiralled down rapidly to the runway. Having landed, crews left two engines running, lowered their cargo ramps to a point just clear of the ground and unloaded their cargoes while taxying slowly through a designated area. With that accomplished, full power was applied and the aircraft took off, climbing steeply for a safe altitude. The tactic minimized time on the ground where transports were most vulnerable to hostile fire.

The formal end of the Bosnian civil war has not brought peace to the Balkans nor an end to NATO intervention. In August 1997 six CF-18s were dispatched to Aviano, Italy, to support a NATO Stabilization Force in Bosnia. They flew air patrols to enforce a 'no fly' zone aimed at preventing air attacks by any side on civilians and peacekeeping ground forces. The Hornets returned to Canada after two months, but in June 1998 another six CF-18s flew to Aviano, prepared to support a possible NATO intervention in Kosovo. It was not an impressive force, although augmentation was possible in the event of a more interventionist NATO policy demanding a larger Canadian role.

* * *

The Persian Gulf War was akin to the Korean War — an American expeditionary force sanctioned by United Nations resolutions and supported by token forces from other nations, each contributing for its own political reasons. The Canadian commitment was almost wholly sea and air in composition; ground forces were committed only to guard Canadian air bases and provide medical support to other armies. This reflected Canadian moods; a poll taken in January 1991 indicated that fifty-three per cent of Canadians wanted their troops to play only defensive roles in the area. However, by late February Canadian support for the war had changed; fifty-eight per cent of the populace favoured war against Iraq and thirty-eight per cent opposed such a move.[11]

In October 1990 one squadron of CF-18s was deployed to Qatar. A crew rotation followed in December; personnel of 439 and 416 squadrons now flew the aircraft, the number of which was increased to twenty-four. The unit, known as the 'Desert Cats' (derived from its formal name, Canadian Air Task Group Middle East — or CATGME), was thus a composite outfit.

Initially their role was purely defensive, in keeping with the concept of DESERT SHIELD, protecting the ships that blocked Iraqi-controlled ports. On the night of 16/17 January 1991, Coalition forces began an intensive air campaign directed at Iraq's command and control systems; DESERT SHIELD now became DESERT STORM. Canadian pilots patrolled in search of Iraqi fighters and escorted aircraft engaged in the offensive. On 30 January 1991 two CF-18s strafed an Iraqi patrol boat; an attempt to hit it with missiles failed because the target gave off too small a heat signature for missile guidance systems to take effect. On 24 February 1991 they commenced ground attack work just as the Coalition commenced a land offensive that lasted merely 100 hours before a ceasefire took effect. Throughout DESERT SHIELD and DESERT STORM the Desert Cats flew 2,700 hours and undertook fifty-six bombing sorties using weapons released under radar guidance. It was the first time since 1945 that Canadian aircraft had been deliberately placed in harm's way, and the task was completed without losing an aircraft.

Apart from the units moved to the area, Canadians served with Allied forces. As of August 1990, when the Gulf crisis erupted, forty-two Canadian air force personnel were on duty at 552 AWACS Wing, Tinker Air Force Base, Oklahoma, and several of these went to Saudi Arabia. The AWACS directed fighters intercepting the few Iraqi aircraft that took off, and thirty-six hostile aircraft were destroyed, all but one case involving AWACS crews that included at least one Canadian. In addition, up to thirteen Canadians on exchange duties were serving with British air forces at any given time and five exchange officers were with American forces.[12]

* * *

The Gulf War saw Canada's air forces committed to battle for the first time in thirty-eight years, yet it came at a moment when the very need for such forces was to become a central question in Canadian decision-making. In 1988–89 the Soviet Bloc had crumbled from within; the fall of the Berlin Wall in 1989 was symbolic, the disbandment of the Warsaw Pact in 1990 more significant. With the Cold War enemy disappearing, NATO itself seemed close to becoming redundant. What nation now posed a threat warranting extensive military preparations? The Canadian government saw none in prospect; the NATO contingents (air and land) were reduced, then withdrawn to Canada, and a search began for a new raison d'être for the forces. The situation was not unlike that which had existed in 1920–21 and again in 1946–47. In those instances, answers had finally been provided: by the rise of Hitler in 1933–34 and the evolution of the Cold War in 1947–49. No comparable clear-cut solutions have presented themselves since 1991.

Certainly the forces have sought earnestly for a role in the New World Order, which is neither as peaceful nor as stable as the old order that existed prior to 1990. Some of the most advanced and sophisticated technology has also been the most difficult to justify. The use of CF-18s to track an aircraft engaged in drug smuggling (November 1992) provided some favourable press coverage, yet the incident cited was inconclusive. Fighters trailed a Convair transport, but it refused to land when instructed and the CF-18s were not authorized to shoot it down. They had to break contact to refuel and the Convair was lost. It was rediscovered by CAF helicopters at an airstrip northeast of Montreal and the smugglers fled when four CF-18s engaged in an unrelated training exercise overflew the field. The culprits were eventually arrested but with no help from the costly fighters. More recently, the inability of CF-18s to shoot down an errant weather balloon in September 1998, described in one report as 'like trying to shoot down a cloud,' generated negative publicity.

These well-publicized setbacks mask unpretentious but solid achievements. A drug seizure at Clova, Quebec, on 11 September 1996 was the climax of a four-day operation involving CF-18s, Auroras and Griffons tracking an aerial smuggler from Colombia to Canada. In domestic crises such as the Saguenay floods (1996), Red River floods (1997) and Quebec-Ontario ice storm (1998), ground troops have been much in evidence; their critically important aerial backup teams have not.

Armed forces are never the same in reality as they appear to the public, but these divergent perceptions have never been so great as they are in the late twentieth century. The popular view of air forces is that of aircrews and their equipment engaged in vigorous training and operations. Increasingly, however, modern training has involved complex simulators, not just for aircraft but for tanks and ships as well, so that intensive instruction can be given in situations where it would be unwise to risk costly machines and precious crews. These have been developed by competing high-tech firms; currently, for example, Canadian Marconi is marketing flight simulators for eight types of aircraft, including the Griffon helicopter and Hercules transport.[13]

Equipment procurement has always had a political edge (the Ross rifle was an obvious example in the first decade of this century) but it has been more clearly politicized in recent years. The process of helicopter purchases in 1992–93 was particularly striking; a Progressive Conservative government decision to buy thirty-five (later twenty-eight) EH-101 helicopters was cancelled, following the 1993 election, by a succeeding Liberal government. The wisdom of this move was questioned at the time; its full impact became apparent in 1998 when an order was finally placed for similar machines to do SAR work without addressing the fleet's expensive ship-borne anti-submarine needs. In August of that year the Navy held a party to observe the thirty-fifth anniversary of the Sea King helicopter in Canadian service, although these craft now spent twenty-five hours in maintenance shops for every hour in the air. The crash of a SAR Labrador helicopter on 2 October 1998, killing all six crewmen, underlined the desperate condition of the SAR fleet; the type was grounded for several weeks, save for only the most dire emergency calls.

The 1994 *White Paper on Defence* marked a turning point for Canada's air forces. The emphasis was now on army needs; land and air elements each had about $3 billion in budgets and by 1998 the air arm budget had fallen to $2.2 billion. It was expected that by the turn of the century the air forces would employ 13,400 men and women, down from 19,500 in 1994–95. The service will have sixty CF-18s, twenty-one modern maritime patrol aircraft (though some with minimal radar equipment) and thirty CC-130 transports. The T-33 combat training fleet will have declined from forty-nine to twenty-five, while the inventory of Challenger aircraft used in coastal surveillance, transport, electronic warfare and other training will have fallen from fourteen to eight machines.

On the other hand, in keeping with the emphasis on army needs, the forces will have ninety-nine Bell 412 helicopters for tactical transport and the first (delayed) deliveries of Cormorant SAR machines. Plans are afoot to upgrade the CF-18 and CP-140 aircraft. New equipment will include Global Positioning System (GPS) receivers — the most advanced navigation devices in the world — in all Canadian Forces aircraft. Investigations to replace the CC-130 Hercules have begun, as has consideration of acquiring aircraft to provide high-speed in-flight refuelling (a capability lost with the retirement of the Boeing 707s).

'Privatization' has also affected the forces. The NATO Flying Training programme in Canada is now a joint venture with Bombardier Services. The scheme has been marketed abroad as a means of reducing training costs, and Denmark has signed a twenty-year agreement incorporating six students annually into the Canadian training programme. The Royal Air Force is expected to participate and Singapore has expressed interest.[14]

The Canadian Forces, air elements included, now face challenges with few precedents, and the observation 'What's past is prologue' may no longer be valid. The end of each world war promised a lasting peace, yet each was followed by major power tensions and wars, hot and cold. The future holds no assurance of similar events. For what threats must nations now prepare? Terrorism? Criminal elements? Ethnic conflicts? Natural disasters, perhaps aggravated by global ecological degradation? And what will be the appropriate weapons to deal with these new dangers? A supersonic fighter is scarcely credible as a riposte to a zealot intent on introducing poison gas into a subway system, and one cannot expect heavily armed maritime patrol bombers to attack foreign trawlers encroaching on national fishing grounds. Although the Balkan and Persian Gulf conflicts suggest that traditional weapons used in projecting power abroad are not yet finished, do Canadians wish to participate in such displays of force? These questions are easy to ask, but the only certainty is that tomorrow is best met with abundant knowledge, flexibility and common sense. History may warn and instruct, even if it can no longer guide.

NOTES

CHAPTER I

1. Unless otherwise specified, all quotations in this chapter are from the first volume of the Official History of the RCAF, S.F. Wise's *Canadian Airmen and the First World War* (Toronto: University of Toronto Press, 1980).
2. H.G. Anderson, *The Medical and Surgical Aspects of Aviation* (London: Frowde and Hodder and Stoughton, 1919), 43.
3. F.H. Hitchins, 'The Canadian Aviation Corps,' in *The CAHS Journal*, Vol 15, No 1 (Spring 1977).
4. H.A. Farr, 'Canada's Air Force and the World War,' in *Canadian Air Review*, Vol 1, No 5 (July 1928).
5. Quoted in J.D.F. Kealy and E.C. Russell, *A History of Canadian Naval Aviation* (Ottawa: Queen's Printer, 1965), 2.

CHAPTER II

1. Canada, House of Commons, *Hansard*, 1919 Session, Vol 2, 1864–65.
2. Air Board report, 1919, quoted in F.H. Hitchins, *Air Board: Canadian Air Force and Royal Canadian Air Force* [Canadian War Museum Paper No 2, Mercury Series] (Ottawa: Canadian War Museum, 1972), 9.
3. Lawrence biographical file, DHist.
4. W.A.B. Douglas, *The Creation of a National Air Force* (Toronto: University of Toronto Press, 1986), 91.
5. T.A. Lawrence, 'The Hudson Strait Expedition,' in *CAHS Journal*, Vol 20, No 3 (Fall 1982).
6. Lewis, 'Adrift on Ice-floes,' DHist 181.001 (D6).
7. Quoted in Hitchins, *Air Board*, 203–4.
8. F.M. Gobeil, 'Siskin Pilot', in *CAHS Journal*, Vol 15, No 1 (Spring 1977), 6.
9. Quoted in Douglas, *The Creation of a National Air Force*, 122.
10. *Hansard*, 1929, Vol 3 (debates of 4 and 14 June).
11. W.C. Clements, 'Preventive Operations,' in *CAHS Journal*, Vol 17, No 1 (Spring 1979).
12. *Hansard*, 1934 Session, Vol 2, 1529.
13. Ibid., 1550.
14. 'Wapitis,' Dunlap biographical file, DHist.
15. Croil to MDN, 2 December 1938, in McGill Papers, DHist 74/628, file A4.
16. L.J. Birchall, 'Trenton to Darmouth,' in *CAHS Journal*, Vol 23, No 2 (June 1985).
17. Douglas, *The Creation of a National Air Force*, 143.
18. Air Force General Order 2/38.
19. *Hansard*, 1939 Session, Vol 3, 3266.
20. *Hansard*, 1939 Session, Vol 2, 2147.

CHAPTER III

1. L.B. Pearson, *Mike: The Memoirs of the Right Honorable Lester B. Pearson* (Toronto: University of Toronto Press, 1972), 208.
2. Quoted in W.A.B. Douglas, *The Creation of a National Air Force* (Toronto: University of Toronto Press, 1986), 196.
3. King diary, 28 May 1948, in National Archives of Canada, MG 26.
4. Canada, Department of External Affairs, *Documents on Canadian External Relations*, Vol 7 (Ottawa: Department of External Affairs, 1972), 643.
5. Ibid., 637.
6. *Souris Plaindealer*, 15 January 1941.
7. Quoted in Spencer Dunmore, *Wings for Victory* (Toronto: McClelland and Stewart, 1994), 97.
8. Ibid., 109.
9. Quoted in F.J. Hatch, *Aerodrome of Democracy: Canada and the British Commonwealth Air Training Plan, 1939–1945* (Ottawa: Department of National Defence, Directorate of History, 1983), 144.
10. Ibid., 141.
11. Dunmore, *Wings for Victory*, 146–47.
12. Douglas Harvey, *Boys, Bombs and Brussels Sprouts* (Toronto: McClelland and Stewart, 1981), 28.
13. Sir Arthur Harris, *Bomber Offensive* (New York: Collins, 1947), 96–97.
14. Hatch, *Aerodrome of Democracy*, 101.
15. Norman Ward, ed., *The Memoirs of Chubby Power* (Toronto: Macmillan, 1966), 224.
16. Harvey, *Boys, Bombs and Brussels Sprouts*, 20.
17. Portal to Breadner, 12 August 1943, in NAC, RG 24, Vol. 5366.
18. *Hansard*, 16 February 1944, 535.

CHAPTER IV

1. Quoted in W.A.B. Douglas, *The Creation of a National Air Force* (Toronto: University of Toronto Press, 1986). All other quotations in this chapter, unless otherwise identified, are from Brereton Greenhous et al., *The Crucible of War, 1939–1945* (Toronto: University of Toronto Press, 1994).
2. Kim Abbott, *Gathering of Demons* (Perth, Ont.: Inkerman House, 1986), 93, 95.
3. H.C. Godefroy, *Lucky Thirteen* (Stittsville, Ont.: Canada's Wings, 1983), 71.
4. Jerrold Morris, *Canadian Artists and Airmen, 1940–45* (Toronto: The Morris Gallery, nd), 65–66.
5. Douglas Alcorn with Raymond Souster, *From Hell to Breakfast* (Toronto: Intruder Press, 1980), 80–81.

CHAPTER V

1. W.A.B. Douglas, *The Creation of a National Air Force* (Toronto: University of Toronto Press, 1986), 493. Unless otherwise identified, all quotations in this chapter are from this book or from Brereton Greenhous et al., *The Crucible of War, 1939–1945* (Toronto: University of Toronto Press, 1994).

² See Stephen L. McFarland and Wesley Phillips Newton, *To Command the Sky: The Battle for Air Superiority over Germany, 1942–1944* (Washington: Smithsonian Institution Press, 1991), 135–36.

CHAPTER VI

1. Brereton Greenhous, et al., *The Crucible of War, 1939–1945* (Toronto: University of Toronto Press, 1994), 402–3. Unless otherwise identified, all quotations in this chapter are from the above volume.
2. Bill McRae, 'The Wingmen', in *The Observair* (Ottawa Chapter Newsletter, CAHS), Vol 35, No 8.
3. No author, *The R.C.A.F. Overseas: The Sixth Year* (Toronto: University of Toronto Press, 1949), 238.
4. No author, *The R.C.A.F. Overseas: The Fifth Year* (Toronto: University of Toronto Press, 1945), 188.
5. In his foreword to David Irving's *The Destruction of Dresden* (London: William Kimber, 1963), 5.
6. Anonymous, *Arnhem Lift* (London: np, 195), 62–63.

CHAPTER VII

1. See Herbert Fairlie Wood and John Swettenham, *Silent Witnesses* (Toronto: Hakkert, 1974).
2. *DND Report (1949)*, 15. See Also 'Armed Forces Fight Floods,' *Canadian Army Journal*, May 1948, 26–28; 'Operation Flood', *Canadian Army Journal*, June 1948, 27–30.
3. Robert McIntyre, *CF-104 Starfighter* (Ottawa: SMG Publishing, 1984), 8. See Also David L. Bashow, *Starfighter: A Loving Retrospective of the CF-104 Era in Canadian Fighter Aviation, 1961–1986* (Toronto: Fortress Publications, 1990).
4. *Minutes of Proceedings and Evidence, Special Committee on Defence,* 14 November 1963, 700.
5. James Eayrs, *In Defence of Canada: Peacemaking and Deterrence* (Toronto: University of Toronto Press, 1972), 336.
6. Desmond Morton, *A Military History of Canada: From Champlain to the Gulf War* (Toronto: McClelland and Stewart, 1992), 242. See also Joseph T. Jockel, *No Boundaries Upstairs: Canada, the United States and the Origins of North American Air Defence, 1945–1958* (Vancouver: UBC Press, 1987).

CHAPTER VIII

1. Carl Mills, *Banshees in the Royal Canadian Navy* (Willowdale, Ont.: Banshee Publications, 1991).
2. T.G. Coughlin, 'Khaki in the Blue,' *Roundel*, October 1964.
3. See *That Others May Live: 50 Years of Para-Rescue in Canada* (Para-Rescue Association of Canada, 1994) for details of numerous mercy missions.
4. *Defence 1974*, 35.
5. Colin A. Morrison, *Voyage into the Unknown: The Search and Recovery of Cosmos 954* (Stittsville, Ont.: Canada's Wings, 1983).
6. Robert McIntyre, *CF-101 Voodoo* (Ottawa: SMG Publishing, 1984), 40.
7. *Report of the Special Committee of the Senate on National Defence: Canada's Territorial Air Defence* (Ottawa: Supply and Services Canada, 1985), 5.
8. Ibid., 7–10. See also *A History of the Air Defence of Canada* (Ottawa: NBC Group, 1997).
9. David L. Bashow, *Starfighter: A Loving Retrospective of the CF-104 Era in Canadian Fighter Aviation, 1961–1986* (Stoney Creek, Ont.: Fortress Publications, 1990), 61.
10. Ibid., 64.
11. J.L. Granatstein and David Bercusson, *War and Peacekeeping: From South Africa to the Gulf — Canada's Limited Wars* (Toronto: Key Porter Books, 1991), 247–248.
12. *Defence 1990*, 12–13; David N. Deere (ed.), *Desert Cats: The Canadian Fighter Squadron in the Gulf War* (Stoney Creek, Ont.: Fortress Publications, 1991); Jean H. Morin and Richard H. Gimblett, *Operation Friction: The Canadian Forces in the Persian Gulf* (Toronto: Dundurn Press, 1997); Desmond Morton, *A Military History of Canada: From Champlain to the Gulf War* (Toronto: McClelland ans Stewart, 1992), 270–271.
13. Christopher Guly, 'The Human Factor: Marconi Simulators Let Aircrews Test Themselves,' Ottawa *Citizen,* 7 October 1998.
14. Sharon Hobson, 'Upgrades in Store for Air Forces,' Ottawa *Citizen,* 7 October 1998.